PRACTICAL X-RAY SPECTROMETRY

PRACTICAL X-RAY SPECTROMETRY

R. JENKINS

J. L. DE VRIES

SECOND EDITION

PHILIPS TECHNICAL LIBRARY

SPRINGER—VERLAG NEW YORK INC.

Sole distributor in the United States and Canada,
Springer-Verlag New York Inc.

U.D.C. No. 543.422.8
ISBN-13: 978-1-4684-6284-5   e-ISBN: 978-1-4684-6282-1
DOI: 10.1007/978-1-4684-6282-1
Library of Congress Catalog Card Number: 72-113623

PHILIPS

Trademarks of N.V. Philips' Gloeilampenfabrieken

First edition 1967
Second impression 1968
Second edition 1969
Second impression 1972
Third impression 1973

PREFACE

X-ray fluorescence spectrometry is now widely accepted as a highly versatile and potentially accurate method of instrumental elemental analysis and so it is somewhat surprising that although the volume of published work dealing with the technique is high the number of textbooks dealing exclusively with its application is relatively few. Without wishing to detract from the excellence of the textbooks which are already available we have both felt for some time, that a great need exists for a book dealing with the more practical aspects of the subject. For a number of years we have been associated with the provision and arrangement of X-ray schools for the training of new X-ray spectroscopists as well as in the organisation of conferences and symposia whose aims have been to keep the more experienced workers abreast with the latest developments in instrumentation and techniques. In all of these ventures we have found a considerable dearth of reference work dealing with the reasons why an X-ray method has not succeeded as opposed to the multitude of success stories which regularly saturate the scientific press.

In this book, which is based on lecture notes from well established courses in X-ray fluorescence spectrometry, we have tried to cover all of the more usual practical difficulties experienced in the application of the method and we have endeavoured to keep the amount of purely theoretical data at a minimum. Wherever possible we have used worked examples to illustrate specific points, particularly in the sections on counting statistics and quantitative analysis. In addition to chapters dealing with the more obvious major headings such as dispersion and detection we have chosen to devote whole chapters to topics such as pulse height selection and sample preparation, lack of experience of which can make all the difference between the accuracy of a method being barely sufficient and comfortably adequate.

It is inevitable that in a rapidly expanding field such as this any published work whose time of preparation is somewhat protracted has specific deficiencies and this book is certainly no exception. The rapid expansion of the spectral range into the ultra-soft X-ray and vacuum ultra-violet region is a case in point and one in which we have had to rigorously restrict our long wavelength limit to about 15 Å. The latest work on wavelength shift and absorption edge fine structure is also covered only sketchily and it may be as well to point out that these areas of expansion of the technique could well become of paramount interest to the analytical chemist during the course of the next few years.

Those new to X-ray fluorescence analysis should find little difficulty in keeping up to date in this field since several literature reviews appear regularly In addition to the excellent general reviews appearing biannually in 'Analytical Chemistry', our own company has for some years now provided quite comprehensive literature reviews dealing exclusively with X-ray analysis. As we have relied heavily on both of these sources of literature we would like to express our thanks to those concerned with their preparation. As the number of references quoted in this book is rather large, we have marked with an asterisk those which we consider to be the more useful. We hope that this will help to guide the reader in his choice for further study.

Since we both started our professional careers as chemists we are particularly aware of our shortcomings in the field of pure physics and we apologise to the physicist, on whose preserves we have encroached, for stating our arguments in terms which we hope are more intelligible to the analytical chemist. We are particularly grateful to our assistants and to Mr. P. W. Hurley for their many useful comments and advice and to the large number of scientists who have passed through our laboratories for providing us with a wealth of interesting problems and fascinating studies.

R. Jenkins
J. L. de Vries

PREFACE TO THE SECOND EDITION

We are very pleased that already now, two years after the first edition, followed by a second impression in 1968, a second edition was found to be necessary.

We have taken the opportunity to revise the text thoroughly to bring it entirely up-to-date with the latest developments in this field. A new section has been introduced dealing in detail with the dead time. In the sections Quantitative Analysis, Mathematical Corrections and Pulse Height Selections major revisions have been made.

We hope that thereby the new book will be even more useful than the first one.

THE AUTHORS

CONTENTS

CHAPTER 1

PHYSICS OF X-RAYS

1.1 Origin of X-rays

1.1.1 GENERAL

The X-ray region is normally considered to be that part of the electromagnetic spectrum lying between 0.1-100 Å, being bounded by the γ-ray region to the short wavelength side and the vacuum ultra-violet region to the long wavelength side. The actual boundary between the X-ray and vacuum ultra-violet region is not clearly defined and for many years the 50-500 Å mid-region has not been exploited by practical spectroscopists to any great degree. Over the last few years however this wavelength range has been examined both from the short wavelength end by the X-ray spectroscopist and from the long wavelength end by workers in the fields of plasma- and astrophysics. It is now common practice to refer to this particular region as the soft X-ray and vacuum ultra-violet region.

Following the discovery of X-rays in 1895 by Röntgen[1], many experiments were made which demonstrated the dual nature of X-rays[2]. Scattering and ionisation experiments indicated the corpuscular nature of the radiation whilst wave character was confirmed, after some difficulty, by diffraction experiments using at first very fine slits and later as in the classic work of Von Laue[3] by use of crystals.

When an element is bombarded with electrons the spectrum obtained in the X-ray region is similar to that shown in Fig. 1.1. This illustrates the main features of the spectrum obtained from a tungsten anode X-ray tube operating at 100 kV and it will be immediately obvious that the spectrum consists of a broad band of continuous (white) radiation superimposed on top of which are discrete wavelengths of varying intensity. The origin of X-ray spectra can be satisfactorily explained in terms of the Bohr concept of the atom, and indeed it was due almost exclusively to X-ray measurements in the early part of this century, that this concept was accepted.

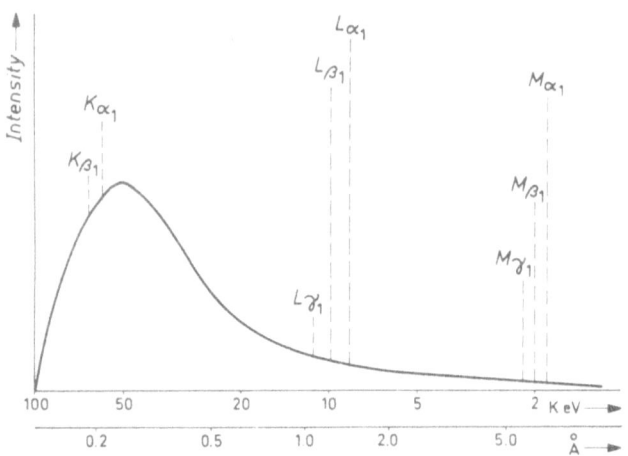

Fig. 1.1 Intensity distribution from a tungsten anode X-ray tube at 100 kV.

1.1.2 CONTINUOUS RADIATION

Continuous radiation occurs following deceleration of the exciting electrons, due to interaction with the impinging electrons and those of the target element. The intensity distribution of this continuum i.e. the number of photons as a function of their respective energy, is characterised by a short wavelength limit λ_{min}, corresponding to the maximum energy of the exciting electrons and by a peak maximum approximating to $2\lambda_{min}$[4]. The short wavelength limit was first defined by Duane and Hunt[5] who demonstrated that this is inversely proportional to the applied potential (V_0), in fact

$$\lambda_{min} = \frac{hc}{V_0} \tag{1.1}$$

where h is Planck's constant and c the velocity of light. λ is expressed in Ångstroms and V_0 in kilovolts. Substitution of these values gives

$$\lambda_{min} \simeq \frac{12.4}{V_0} \tag{1.2}$$

Although the overall intensity of the continuum increases with applied current (i) and potential (V) and also with the atomic number (Z) of the target material, the relative distribution remains virtually constant. However, as will be seen later, the spectral distribution from an X-ray tube designed for spectrometry can vary significantly at the long wavelength end due to inherent window filtration. There have been many attempts to express the distribution of the continuum in terms of excitation conditions and probably one of the most useful of these is that due to Kramers.[6]

$$I\lambda \cdot d\lambda = K \cdot i \cdot Z \left| \frac{\lambda}{\lambda_{min}} - 1 \right| \frac{1}{\lambda^2} \cdot d\lambda \qquad (1.3)$$

This formula relates the intensity $I(\lambda)$ from an infinitely thick target of atomic number (Z) at any wavelength (λ) with the applied current (i). λ_{min} has already been defined in Equation (1.2) and K is a constant. This expression does not correct for self absorption by the target — a factor which becomes increasingly more significant at the long wavelength end of the distribution. Since λ_{min} is inversely proportional to the applied potential it will be seen that the intensity of the continuum is a roughly linear function of the tube current and the atomic number of the target material. It is, however, a rather more complex function of the applied potential. The significance of these facts will become more apparent later in the discussion of the production of X-rays and design of X-ray tubes.

1.1.3 CHARACTERISTIC RADIATION

Characteristic radiation arises from the energy transferences involved in the re-arrangement of orbital electrons of the target element following ejection of one or more electrons in the excitation process. Fig. 1.2 illustrates

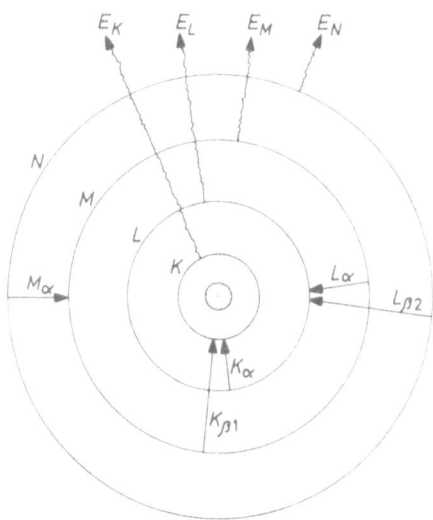

Fig. 1.2 Transitions giving X-radiation.
$(E)K_\alpha = E_K - E_L$
$(E)K_{\beta 1} = E_K - E_M$
$(E)L_\alpha = E_L - E_M$
$(E)L_{\beta 2} = E_L - E_N$
$(E)M_\alpha = E_M - E_N$

this concept and indicates the more important of the transitions which are involved. For example, if an electron from the K shell is ejected (i.e. removed to infinity as far as that atom is involved which may simply mean promotion to a conduction band) the atom becomes unstable due to the presence of a 'positive hole' in the K shell. The stability of the atom can however be regained by single or multiple transitions from outer orbitals since in general, the potential instability due to ionisation of an atom decreases in the order $K^+ > L^+ > M^+ > N^+$ etc. Each time an electron is transferred the atom moves to a less energetic state and radiation is emitted at a wavelength corresponding to the difference in the energy between the initial and final states of the transferred electron. The energy of the transferred electron will correspond to the potential required for its removal from its particular shell. For instance, if an electron is first ejected from the K shell and this hole is filled by an L electron the energy associated with this transference will be equivalent to $(E_K - E_L)$, in fact the Kα line. The hole in the L shell may then be filled by an M electron with the emission of an L line of energy $(E_L - E_M)$. This process will continue until the energy of the atom is lowered to a value approximating to that associated with normal electron vibration in the outer orbitals — in general a few electron volts.

In practice the processes involved are rather more complex than the concept described above might suggest and this is due to the fact that each electron in an atom can have energy other than that due to its position in a certain shell (i.e. its principal quantum number). As is well known the energy of any electron is defined by four separate factors n, l, m, and s, called the quantum numbers of the electron. n is the principal quantum number and can take on integral values 1, 2, 3 etc. The K shell has $n = 1$, the L shell $n = 2$ and the M shell $n = 3$, etc. l is the angular quantum number which determines the shape of the orbital. Each orbital can in turn hold up to two electrons. l can have values of $(n - 1) \ldots 0$ and when $l = 0$ the orbital is called an 's' orbital, when $l = 1$ a 'p' orbital, $l = 2$ a 'd' orbital and $l = 3$ an 'f' orbital. m the magnetic quantum number is the projection of the angular momentum defined by l upon the direction of the magnetic field and this can take on all values of $+l$, $-l$ or 0. Finally s, the spin quantum number can take on values of $\pm \frac{1}{2}$. The Pauli exclusion principle states, that no two electrons in the same atom can have the same set of quantum numbers, and hence it is possible to predict the maximum number of electrons in each shell to $2n^2$. Table 1.1 lists the atomic structures of the first three shells.

Since we are considering the total energy change associated with the change in the principal quantum number it is apparent that effects due to changes in the other quantum numbers must also be considered. To a first

TABLE 1.1

Atomic structures of first three principle shells

Shell	n	l	m	s	Maximum number of electrons	Possible values of J
K	1	0	0	$\pm\frac{1}{2}$	2	$\frac{1}{2}$
L	2	0	0	$\pm\frac{1}{2}$	8	$\frac{1}{2}$
		1	+1	$\pm\frac{1}{2}$		$\frac{1}{2}, \frac{3}{2}$
		1	0	$\pm\frac{1}{2}$		
		1	−1	$\pm\frac{1}{2}$		
M	3	0	0	$\pm\frac{1}{2}$	18	$\frac{1}{2}$
		1	+1	$\pm\frac{1}{2}$		$\frac{1}{2}, \frac{3}{2}$
		1	0	$\pm\frac{1}{2}$		
		1	−1	$\pm\frac{1}{2}$		
		2	+2	$\pm\frac{1}{2}$		$\frac{3}{2}, \frac{5}{2}$
		2	+1	$\pm\frac{1}{2}$		
		2	0	$\pm\frac{1}{2}$		
		2	−1	$\pm\frac{1}{2}$		
		2	−2	$\pm\frac{1}{2}$		

approximation the energy of an electron is determined by its configuration which in turn is dependent only upon n and l. The influence of the spin quantum number however is sufficiently large to confer significant changes in l and it is necessary to consider the vector sum of l and s. This vector sum is called J and is the projection on the direction of the magnetic field, i.e.

$$\bar{J} = l + \bar{s} \tag{1.4}$$

Fig. 1.3 illustrates this effect by means of a vector diagram and demonstrates that two values of J are possible for an electron with $l = 1$. This statement is true for all values of l greater than or equal to 1, but where $l = 0$ the orbital is spherically symmetrical and only one value of J is possible i.e. $J = +\frac{1}{2}$. As will be seen from Fig. 1.3 there is one possible value of J in the K shell, three in the L shell and five in the M shell. For this reason we refer to the existence of one K group, three L groups designated L_I, L_{II} and L_{III} and five M groups designated M_I to M_V.

The selection rules which determine the allowable transitions are that $\Delta l = \pm 1$ and $\Delta J = 0$ or ± 1. Fig. 1.4 illustrates the characteristic lines which fit these selection rules and also the accepted terminology. The accepted nomenclature of the X-ray lines is unfortunately somewhat unsystematic but in general the final resting place of the transferred electron determines the series or group to which the line belongs. Further to this an α line is always associated with $\Delta n = 1$ and the strongest β and γ lines occur when $\Delta n = 1$ or 2.

Since the wavelength of the radiation is inversely proportional to the difference in energy between the initial and final states of the transferred electron, it necessarily follows that wavelengths of lines within a series will decrease as the energy gap increases. Therefore lines involving $\Delta n = 2$ will be harder (i.e. of shorter wavelength) than those arising from $\Delta n = 1$. Thus the Kβ line is harder than the Kα line and so on. It will also be apparent that the wavelengths of different series will increase from K to L to M etc.

A general expression relating the wavelengths (λ) of a characteristic line

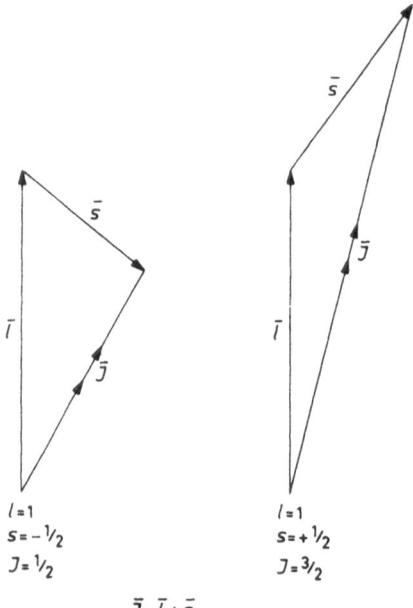

$l = 1$
$s = -\frac{1}{2}$
$J = \frac{1}{2}$

$l = 1$
$s = +\frac{1}{2}$
$J = \frac{3}{2}$

$\bar{J} = \bar{l} + \bar{s}$

Fig. 1.3 Vector sum of angular and spin quantum numbers.

with the atomic number (Z) of the corresponding element is given in Moseley's Law[7] i.e.

$$\frac{1}{\lambda} = k[Z - \sigma]^2 \tag{1.5}$$

where k is a constant which varies with the spectral series and σ is a screening constant which corrects for the repulsion due to other electrons in the atom. Appendix 3 tabulates the wavelengths and origins of the major K and L series lines along with their approximate relative intensities. The intensity of a characteristic line is a function of the transition probability. Once an atom has been ionised in, for example, the K-shell there is a certain probability that this hole will be filled by an electron from the $L_{II}(K\alpha_2)$, $L_{III}(K\alpha_1)$, $M_{III}(K\beta_1)$ shell and so on. The number of possibilities is even greater for filling a hole in the L-shell. The transition probability is an exponential function of the difference in energy states of the electron in the different orbitals. Thus although the intensity ratios of the various characteristic lines is constant for a given atom, they will gradually change with the atomic number Z. For instance, the intensity ratio $K\alpha : K\beta$ is about 5 : 1 for Cu but smaller for the heavier elements (approx. 3 : 1 for Sn) and much larger for the lighter elements (approx. 25 : 1 for Al). Further to this although the

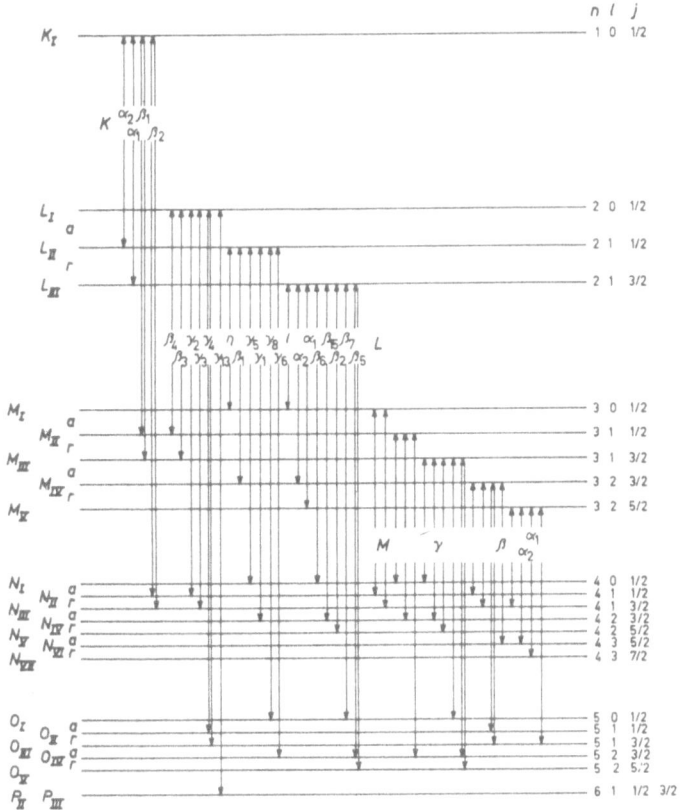

Fig. 1.4 X-ray emission lines.

characteristic wavelengths and their relative intensities are to a first approximation constant for a given element, small changes may occur when the distribution of the outer (valence) electrons changes.

1.1.4 NON-DIAGRAM LINES

A large number of lines have been reported[2] which do not appear to satisfy the selection rules already formulated. The majority of these are very weak and are of little or no consequence to the analytical spectroscopist. Notable exceptions are the non-diagram (satellite) lines which occur in the $K\alpha$ series, the most important of these being the $K\alpha_3$, α_4 doublet. The origin of these lines seems to be a double transition LL—LK [8,9]. An atom can be ionised in the K and L orbital at the same time and the energy connected with a double electron jump may be emitted as a single photon, in this instance the $K\alpha_3$ line, whose wavelength is thus slightly shorter than that of $K\alpha_1, \alpha_2$. The probability of this transition, hence the intensity of the $K\alpha_3$ line, is a function of the relative transition times. The life time of an ionised state is longer for the lighter elements and this explains why for example, the $K\alpha_3$ line is considerably stronger for Al than for heavy and medium heavy elements In fact the intensity for Al $K\alpha_3$ is roughly 10% of that of Al $K\alpha_1, \alpha_2$. The influence of valence state and outer electron distribution on wavelength and intensity is more pronounced in the case of $K\alpha_3$ than for $K\alpha_{1,2}$ or $K\beta$.

1.1.5 AUGER EFFECT

It has already been pointed out that the extra energy which an atom possesses after an electron jump, for example L to K, may be emitted as characteristic radiation. Alternatively, however, this energy may be used to reorganize the electron distribution within the atom itself leading to the ejection of one or more electrons from the outer shell[10]. The probability of this type of ionisation will increase with a decrease in the difference of the corresponding energy states. For example when $(E_K — E_L)$ is only slightly larger than E_L this ionisation probability is large which in turn means that only in a small number of the total original K ionisations the energy is emitted as K radiation. This phenomena is called the Auger effect[11] and is rather akin to the auto-ionisation effect found in optical spectra[12].

1.1.6 FLUORESCENT YIELD

An important consequence of the Auger effect is that the actual number of useful X-ray photons produced from an atom is less than would be expected, since a certain fraction of the absorbed primary photons give rise to Auger electrons. The ratio of the useful X-ray photons arising from a certain shell to the total number of primary photons absorbed in the same shell, is called the fluorescent yield (ω). The value of ω, which is of course necessarily less than unity, decreases markedly with atomic number since the probability of producing an Auger electron increases. Similarly L fluorescent yield values (ω_L) are always less than the corresponding K fluorescent yield values (ω_K). Fig. 1.5 shows approximate curves of ω_K and ω_L. Although the ω_K curve is fairly well established there is a far less equivalent data for ω_L values. This is due to the fact that L fluorescent yields are far more difficult to measure although a fairly comprehensive set of data has recently been obtained using an X-ray coincidence counting method[13]. Study of the ω_K and ω_L curves indicates that for the normal operating range of the conventional X-ray spectrometer i.e. K lines up to atomic number 56, L lines for elements of atomic number greater than 50, fluorescent yield values are less than 0.4 for more than half of the atomic number range. It will also be seen that for wavelengths longer than 3.5 Å fluorescent yield values are less than 0.1 and this is an inherent limitation to the sensitivity of the X-ray method for longer wavelengths.

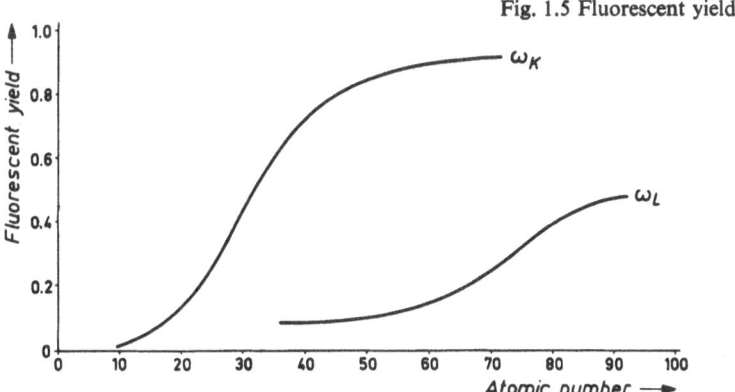

Fig. 1.5 Fluorescent yield.

1.2 Production of X-rays

1.2.1 GENERAL

Much of the early X-ray work was carried out using direct electron excitation but this procedure suffers from the limitations which inevitably arise as a consequence of the need to work under high vacuum conditions. Major problems can arise from sample volatility, local heating and poor conduction and it was limitations such as these that led to the proposal that primary X-rays be excited from an ideal sample (i.e. the target of an X-ray tube) and these primary X-rays then used to excite secondary (fluorescent) radiation from the sample[14]. In consequence practically all conventional X-ray spectrometry is now based on the fluorescence technique although several attempts have been made to exploit the potentially more efficient direct electron excitation method for the lighter elements.[15-16] In addition the advent of the electron microprobe has opened up a vast new field of analysis of very small areas (of the order of square microns as opposed to square centimetres in X-ray fluorescence spectrometry) using focused electron beams. e.g.[17] The use of radio isotopes as a source of primary radiation has also been exploited to a significant degree[18-20] but in general intensities are several orders of magnitude lower than those obtainable with conventional X-ray tubes.

1.2.2 X-RAY TUBES

Figure 1.6 shows a diagrammatic representation of the conventional X-ray tube. The tungsten filament (a) is heated by means of a current (the filament current) producing a region of high electron density around the filament. Part of this electron cloud is accelerated along the anode focusing tube (c) by means of a large potential difference (the tube high voltage), applied between the anode and the filament. Electrons striking the anode (d) produce X-radiation a significant portion of which passes through the window (e). The purpose of the cathode cap (b) is to absorb the unused and scattered electrons and to cut down the spread of tungsten which vaporises from the filament.

Although the construction of an X-ray tube may appear relatively simple, certain design considerations such as choice of anode material and window

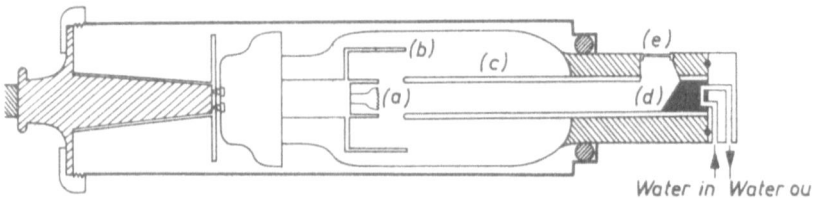

Fig. 1.6 *The sealed X-ray tube.* The tungsten filament *a* is heated by the filament current producing a cloud of electrons. These electrons are accelerated along the focusing tube *c*, by the potential difference between the filament and the anode *d*. The generated X-rays (or a significant part of the total) then pass through the window *e* to the outside.

characteristics can be critical. It will be apparent from equation (1.3) that the strongest continuous radiation can be obtained from anode materials of high atomic number. However the conversion of the high voltage electrons into X-rays is a very inefficient process and only about 1 % of the total applied power emerges as useful radiation. The majority of the remaining energy appears as heat which has to be dissipated by cooling the anode. From this point of view it is vital that the anode be a good heat conductor or at least it should be possible to weld it to a watercooled copper block. Heat dissipation problems are by no means confined to the anode since scattered electrons can also raise the temperature of the X-ray tube window to several hundred degrees centigrade. In order to give high transmission of the longer wavelengths the window is normally constructed of beryllium but the thickness of the window is dependent almost exclusively on its ability to dissipate heat. Beryllium is a poor conductor and if too thin a window is used large temperature gradients can form across its diameter which may eventually lead to fracture. Since the total amount of electron back-scatter increases with the atomic number of the anode, window heating problems are more critical for higher atomic number anodes. For example, a 500 μm thick window might be necessary for a tungsten or gold anode X-ray tube whereas a 200 μm thick window would probably suffice for a chromium anode X-ray tube. The possibility of employing a thin window in the X-ray tube is an extremely attractive proposition for long wavelength work because of the increased window transmission for the longer wavelength continuous radiation. For instance, reduction of the window thickness from 1000-300μm increases the transmission for 4 Å radiation from 3 % to 35 % which in effect can mean an increase of 2-3 in excitation efficiency for the lower atomic number elements such as aluminium and silicon.

1.2.3 LIGHT ELEMENT EXCITATION

For the excitation of longer wavelength radiation there is obviously an incentive fc producing an X-ray tube with a thin window and a low atomic number anode. Several successful attempts have been made to construct such a tube[21-22] at least one of which has been produced on a commercial scale[23]. These specially designed light element tubes have anodes of aluminium, copper, etc. and are continuously pumped. Two windows are normally fitted, the first of these being thick and removable from the outside once a rough vacuum has been obtained. The other window remains in position all the time, but this one is ultra-thin, being constructed of perhaps 1 micron polypropylene. In certain cases it may even be possible to work without a window at all[22].

For routine quantitative analysis conventional sealed tubes have the advantage over open continuously pumped tubes in that their output is far more stable. This is because the residual air pressure is an important parameter in photon production and in continuously pumped tubes it is difficult to keep internal pressure constant. From this point of view sealed tubes are invariably more convenient than pumped tubes so long as their intensities are comparable. To date this means wavelengths down to about 9 Å. However, light element sensitivities have increased by almost an order of magnitude during the period 1961-1965 and it is to be expected that further considerable gains will be achieved as instrumentation becomes more sophisticated.

The optimum choice of X-ray tube parameters is illustrated in Appendix 2b where it will be seen that characteristic line intensity increases sharply for operating potentials just in excess of the critical excitation value. By the time twice this value has been attained however the correlation is approximately linear. The given graph applies to molybdenum using a tungsten anode tube where only the white continuous spectrum is active in exciting MoKα. Slightly different curves are obtained when the characteristic tube lines also contribute to the excitation. It may be as well to point out at this stage that the general requirement for routine X-ray analysis is the provision of a tube giving high spectral output over a very large wavelength range. From this point of view a middle order atomic number anode such as silver or rhodium may be the most useful since the electron back-scatter is only moderate and relatively thin windows can be used. In addition to their rather intense white spectrum the characteristic K lines give a high spectral output in the 0.5-0.6 Å region and the L lines in the 4-5 Å region. Alternatively, a dual anode tube can be employed in which one can switch from a heavy element anode to a light element anode by means of an external mechanism[14]. Although this type of arrangement gives the advantages associated with the alternative sources of intense characteristic lines, its window thickness is still determined by the highest atomic number target element used.

1.3 Properties of X-rays

1.3.1 ABSORPTION

The earliest experiments demonstrated that as X-rays transverse matter they are attenuated by an amount dependent upon the thickness and density of the absorbing medium. It was further shown that X-rays of different wavelength are attenuated by varying amounts by the same absorber. If a monochromatic beam of X-rays of wavelength λ_0 and intensity I_0 is incident upon a homogeneous absorber of thickness x, a certain fraction I will pass through the absorber whilst the remainder, $(I_0 - I)$, will be lost by photoelectric absorption or scatter. The fraction of both absorbed and scattered photons is proportional to I_0 but is also dependent on variations in thickness dx, mass dm, or number of atoms dn, encountered by a beam of cross section 1 cm². If the proportionality constant is designated μ with a subscript x, m or n the following relationship will hold:

$$dI_0 = - I_0 \cdot \mu_x dx \tag{1.6}$$
$$dI_0 = - I_0 \cdot \mu_m dm \tag{1.7}$$
$$dI_0 = - I_0 \cdot \mu_n dn \tag{1.8}$$

The coefficients μ_x, μ_m and μ_n are called respectively the linear absorption coefficient, the mass absorption coefficient and the atomic absorption coefficient.
A simple relationship exists between these coefficients:

$$\mu_x = \mu_m \rho = \mu_n \rho \cdot N/A$$

where ρ is density, N Avagadro's number and A atomic weight. The fraction (I) of photons transversing the absorber without being scattered or absorbed can be calculated by integrating dI_0 between the limits 0 and x. Hence by integration of Equation (1.6)

$$\ln I_x - \ln I_0 = - \mu_x \cdot x$$

by substitution of μ_x by $\mu_m \cdot \rho$

$$I = I_0 \cdot \exp(-\mu\rho x) \tag{1.9}$$

which is an expression of Beer's Law.

The mass absorption coefficient is the most useful of the three absorption terms and it is common practice to refer to this simply as μ. The mass absorption coefficient is a function only of the wavelength of the absorbed radiation and the atomic number of the absorbing element and data has been published relating these variables. For example, the data listed in Appendix 1 is taken mainly from the work of Victoreen.[25] Graphical interpolation has been used to find values of μ for all elements for the range of wavelengths commonly used in X-ray spectrometry with the result that some of these data are rather inaccurate, especially in the case of the lighter elements. More accurate values are being published from time to time but to date no complete set is available for all wavelengths and all elements. For wavelengths up to 2.5 Å a set of weighed averages has appeared for certain wavelengths.[26] Victoreen proposes the following relationship:

$$\mu = C\lambda^3 - D\lambda^4 + \sigma_e\frac{Z \cdot N}{A} \tag{1.10}$$

where C and D are both constants for a given atomic number Z. N/A is the number of atoms per gram and σ_e the scattering coefficient per electron. If a plot of mass absorption coefficient against wavelength is prepared sharp discontinuities called absorption edges are found. Fig. 1.7 shows such a plot for tungsten and it will be seen that there is one K edge and three L edges. Unfortunately, very little data is available in the region of the M edges but as indicated on the plot five such edges are observed. The actual positions of the edges vary with atomic number and the single K edge, the three L edges and the five M edges correspond respectively to the one, three and five allowable J values previously described.

The actual mass absorption coefficient is made up of two components the least important of which is the scattering coefficient σ. σ represents the fraction of incident X-rays scattered per cubic centimetre and is wavelength independent. Far more important is the true photoelectric absorption τ which between edges conforms to the relationship:

$$\tau = \frac{K \cdot N}{A} \cdot Z^4\lambda^3 \tag{1.11}$$

where K is a constant and N/A is the number of atoms per gram. Since

$$\mu = \tau + \sigma \tag{1.12}$$

and since τ is invariably much greater than σ, μ is approximately proportional to a cubic power of λ and a fourth power of Z.

The true photoelectric absorption is made up of photoelectric absorption in each of the principle sub-levels of the atom thus

$$\tau(\lambda) = \tau_K(\lambda) + \tau_L(\lambda) + \tau_M(\lambda) + \tau_N(\lambda) \tag{1.13}$$

Each time λ increases to a value in excess of a certain absorption edge wavelength one of the terms in Equation (1.13) drops out resulting in a sharp fall in the mass absorption coefficient value. The rapid increase in τ as the absorption edge is approached from the short wavelength side indicates that there is some resonance effect in the ionisation process and it is to be expected that the most effective wavelengths causing excitation are to be found just to the short wavelength side of the absorption edges.

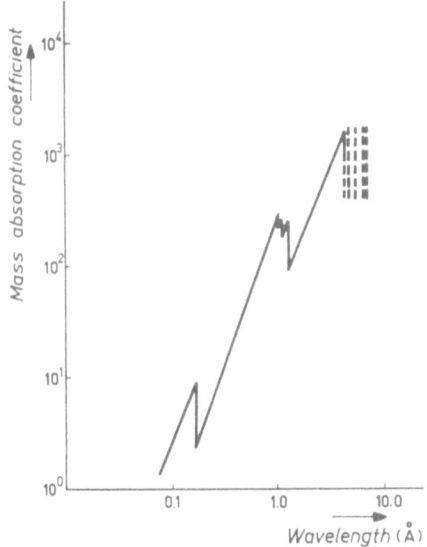

Fig. 1.7 Mass absorption coefficient of tungsten as a function of wavelength.

1.3.2 SCATTERING OF X-RAYS

The scattering coefficient is made up of two terms corresponding to coherent (Rayleigh) scatter and incoherent (Compton) scatter. In fact

$$\sigma = \underset{\text{coherent}}{Zf^2} + \underset{\text{incoherent}}{(1 - f^2)} \tag{1.14}$$

where f is the electronic structure factor.[2]

Coherent scatter arises when an X-ray photon collides with an electron and is deviated without loss of energy, the corresponding wavelength remaining unchanged. If the electron is only loosely bound, the colliding X-ray photon may lose part of its energy to the electron. As the energy of the scattered photon has decreased the scattering process is incoherent. Since the total momentum remains unchanged, it can be shown that the relationship between the incoherent scatter (λ_c) and the incident wavelength (λ_0) is

$$\lambda_c - \lambda_0 = 0.0243(1 - \cos \psi) \tag{1.15}$$

where ψ is the angle through which the radiation is scattered. In the case of the X-ray spectrometer ψ is the angle between the central ray of the primary X-ray beam and the primary collimator axis. In most spectrometers this angle is approximately 90° and since the wavelengths of all coherently scattered lines are unmodified the wavelength difference between coherently and incoherently scattered tube lines is approximately equal to 0.024 Å.

Fig. 1.8 shows a scan over the second order tungsten Lα lines which have been scattered from a tungsten anode X-ray tube using a sample of distilled water. The broad band of incoherently scattered radiation is clearly seen to the long wavelength side of the Lα doublet. As the angle ψ is not constant

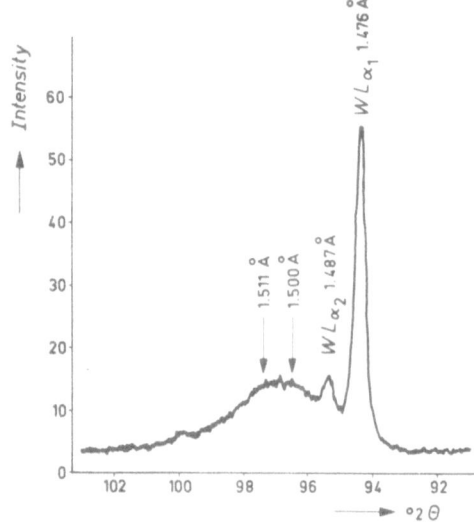

Fig. 1.8 *Effect of Compton scatter on scattered X-ray tube lines.* The spectrum of the second order $WL\,\alpha_1$ and $WL\,\alpha_2$ was recorded using an LiF(200) crystal and a sample of distilled water. The broad Compton peak to the long wavelength side of the coherently scattered lines consists of two incoherently scattered maxima each of 0.024 Å greater than the corresponding coherent line.

for all primary rays, the incoherently scattered peaks are always much broader than the coherently scattered ones. This is because the primary rays do not travel along one direction only, since a cone of approximately 30° aperture irradiates the specimen. This means that wavelength shifts may lie between approximately 0.020 and 0.028 Å. The intensity ratio between the two scattered lines depends on the atomic number Z of the scattering medium and the wavelength λ of the scattered rays. The lower Z and the shorter λ, the higher is the incoherently scattered peak.

1.3.3 ABSORPTION BY COMPOSITE MATERIALS

The mass absorption coefficient of any compound or composite material can be calculated from the relationship

$$\mu(\text{compound}) = \Sigma(\mu_i \cdot W_i) \tag{1.16}$$

where μ_i and W_i are individual mass absorption coefficients and weight fractions. For example, the mass absorption coefficient of KBr for Cu Kα radiation (1.542 Å) is equal to

$$(\mu_{KBr})Cu\ K\alpha = (\mu_K \cdot W_K) \quad + (\mu_{Br} \cdot W_{Br})$$
$$(150 \times 0.328) + (92 \times 0.672)$$
$$= 49.6 \qquad + 61.8$$
$$= 111.4$$

Equation (1.16) can similarly be used for calculating the mass absorption coefficient of any mixture of elements or compounds making up a sample matrix. This value is usually called the "Matrix μ" value for the particular wavelength in question.

1.3.4 DIFFRACTION OF X-RAYS

An electron which is situated in an alternating electromagnetic field will oscillate with the same frequency as the field. Since an X-ray beam can be considered as an electromagnetic wave travelling through space it too will cause all electrons in its path to oscillate. Each electron can then be considered as a small oscillator emitting electromagnetic radiation at the same frequency as the primary radiation, giving the resultant wave of the atom. The amplitude of this wave depends on the number of electron waves and their respective phase differences. These phase differences depend on the differences in path lengths, as illustrated in Fig. 1.9. Let the dots represent

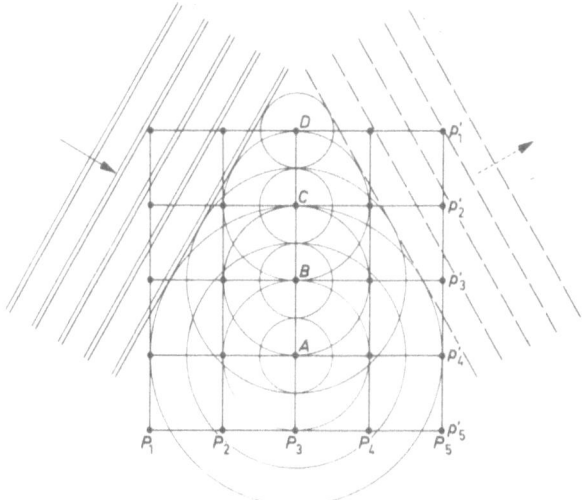

Fig. 1.9 Origin of diffraction phenomena.

scattering centres at a distance d from each other; the circular waves emitted by A, B, C, D when irradiated by a wavefront are indicated. In this case, the wavelength, i.e. the distance between two full lines, the distance d and the angle of incidence, have been chosen such that the electrons oscillate in phase with each other with a phase difference of 1. A resultant wave is formed in the indicated direction, made up of single waves from all of the successive centres with a mutual phase difference of 2, as the wavefront is tangential to the first circle of D, the second of C, the third of B etc. In the direction BP'_3 a similar wave results with the same amplitude and a mutual phase difference of 1. The tangent to the first circle of D, and the second circle of B would give a wavefront with zero intensity as in that direction the phase difference between the waves emitted by centre D and centre C is exactly $1\frac{1}{2}$, which means that these two waves cancel each other; similarly for waves originating at A and B.

An atom consists of a nucleus and many electrons circling around this nucleus in discrete orbitals. When the angle between the impinging radiation and the direction of observation is zero, there will be no phase difference between the waves and the resultant wave will have maximum amplitude. As this angle increases, the waves of the outer opposite electrons will gradually get out of phase. Their contributions then cancel and the amplitude of the resultant wave diminishes. The scattering power f of an atom is thus dependent on its atomic number Z and the direction of observation.

1.3.5 CONDITIONS FOR DIFFRACTION

When a beam of monochromatic X-rays falls onto a crystal lattice, a regular periodic arrangement of atoms, a diffracted beam will only result in certain directions. It is necessary that the waves emitted by the individual atoms be in phase with each other in the direction of observation. Fig. 1.9 illustrates this case for rows of atoms separated by distances d. It is usually more convenient to visualise a crystal lattice as consisting of sets of parallel planes, separated by distances d. All the atoms are situated in these planes.

For instance, the lines P_1P_5 or $P'_5P'_1$ in Fig. 1.9 might be thought of as planes in the two dimensional case. The condition for diffraction can now be found in two steps. First the waves emitted by all atoms lying in a single plane must be in phase, and second the scattering of waves by successive planes must also be in phase. The first condition is fulfilled if the incident ray, the dif-

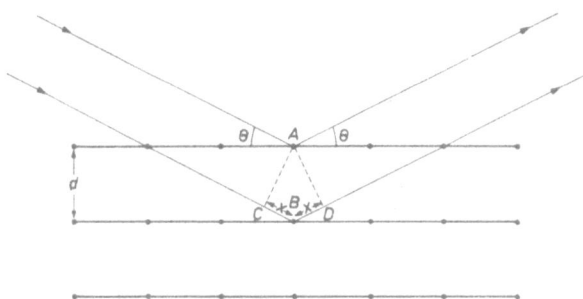

Fig. 1.10 Condition for diffraction of X-rays.

fracted ray and the normal to the reflecting surface all lie in one plane, and if the angle of incidence equals the angle of "reflection". In this case it can be proven that all of the waves scattered by the atoms in this plane are in phase. The second condition is illustrated in Fig. 1.10. Two parallel rays strike a set of crystal planes at an angle θ and are scattered as previously described. Reinforcement will occur when the difference in the path lengths of the two rays is equal to a whole number of wavelengths. This path length difference is equal to $CB + BD$ and since $CB = BD = x$, $n\lambda$ must equal $2x$ for reinforcement, where n is an integer. However, it will be seen that $x = d \cdot \sin \theta$ where d is the interplanar spacing; hence the ultimate reinforcement condition is that

$$n\lambda = 2d \cdot \sin \theta \qquad (1.17)$$

this being a statement of the Bragg Law. Bragg's Law takes no account of the refraction of X-rays but since this effect is very small (the index of refraction is of the order of 0.99999) it can usually be ignored. It can however, become significant when dealing with longer wavelengths where it can be responsible for shifts of about $0.02°\ 2\theta$.

1.4 Excitation of fluorescence radiation in the sample

1.4.1 GENERAL

The excitation of fluorescent radiation within the sample by the polychromatic radiation from the X-ray tube is a complex process and the derivation of any relationship between source intensity and measured fluorescence is necessarily an empirical exercise. Nevertheless such an exercise is useful since it provides a means of establishing the relative importance of the various operational parameters and it is the purpose of the following simplified derivation to indicate the part which the functions already described play in the excitation process. In addition it should provide some of the necessary background information for subsequent chapters dealing with matrix effects and quantitative analysis.

A simple case will first be derived indicating the various stages in the excitation of characteristic photons by monochromatic primary radiation. This will then be extended to include all the wavelengths which make up the polychromatic beam normally employed in X-ray fluorescence spectrometry.

1.4.2 EXCITATION BY MONOCHROMATIC RADIATION

X-ray photons from the X-ray tube enter the sample at an average angle ψ_1. These primary photons will either be scattered (coherently or incoherently) or be absorbed by the atoms of the matrix. The absorbed photons will give rise to photoelectrons from the matrix atoms which will appear either as X-ray photons or Auger electrons. Only the photons emitted in the direction of the collimator can contribute to the measured intensity and these are further absorbed on leaving the sample.

The actual processes involved in the excitation can be conveniently broken down into five separate factors:

(1) The number of primary photons striking the sample surface per unit of time;
(2) The attenuation of this intensity by the absorbing matrix;
(3) The efficiency of the actual excitation of characteristic radiation;
(4) The proportion of these photons which are accepted by the collimator;
(5) The attenuation of the characteristic radiation by the sample matrix.

Table 1.2 lists these factors in terms of defined parameters (see also ref. 30). Let us consider the characteristic radiation of an element j in a sample. The primary intensity $I_0(\lambda)$ making an angle ψ_1 with the sample surface is reduced to $I_x = I_0(\lambda) \exp\left[-\mu(\lambda)\rho_m \cdot x \cdot \mathrm{cosec}\ \psi_1\right]$ reaching a layer of thickness Δx at height x above the surface. In this layer Δx a certain fraction $\Delta I_x = I_x \cdot \mu(\lambda)\rho_m \cdot \Delta x \cdot \mathrm{cosec}\ \psi_1$ is absorbed by all atoms. As previously indicated only the fraction $\dfrac{C_j \cdot \mu_j(\lambda)}{\mu(\lambda)} \cdot \Delta I_x$ is absorbed by the element j.

It has been shown in equations 1.12 and 1.13 that $\tau_K = \mu - \sigma(\tau_L\ G\ \tau_M \ldots)$.

The ratio of the absorption μ just before and just beyond the K absorption edge is called the absorption jump r_j. The actual value of r_j decreases from 9.6 for aluminium to about 2 for tungsten. The fraction of absorbed intensity by element j which leads to K ionisation is thus

$$\Delta I_x \frac{C_j}{\mu(\lambda)} \cdot \frac{r_j - 1}{r_j} \cdot \mu_j(\lambda) = I_x \cdot C_j \cdot \rho_m \cdot \frac{r_j - 1}{r_j} \cdot \mu_j(\lambda) \cdot \Delta_x \cdot \operatorname{cosec} \psi_1$$

Only the fraction $\omega_j \cdot g_j$ leads to Kα radiation thus the intensity ΔI or characteristic Kα radiation emitted by the layer Δx is given by

$$\Delta I_j = I_x \cdot C_j \cdot \rho_m \cdot \frac{r_j - 1}{r_j} \cdot \mu_j(\lambda) \cdot \omega_j \cdot g_j \cdot \Delta x \cdot \operatorname{cosec} \psi_1$$

Only the fraction $\dfrac{d\Omega}{4\pi}$ can pass the collimator, the axis of which makes an angle ψ_2 with the sample surface; this radiation is absorbed passing through the sample and only the fraction $\Delta I_j \cdot \exp.[-\mu(\lambda_j) \cdot \rho_m x \operatorname{cosec} \psi_2]$ reaches the sample surface. The contribution of the layer Δx to the total characteristic intensity of element j is thus given by:

$$\Delta I_j = I_0(\lambda) \exp[-\mu(\lambda)\rho_m x \operatorname{cosec} \psi_1 + \mu(\lambda_j) \cdot \rho_m \cdot x \operatorname{cosec} \psi_2] C_j \cdot \rho_m \cdot$$

$$\frac{r_j - 1}{r_j} u_j(\lambda) \cdot \omega_j \cdot g_j \frac{d\Omega}{4\pi} \Delta x \cdot \operatorname{cosec} \psi_1 \tag{1.18}$$

TABLE 1.2

Factors Making up the Process of Fluorescence Excitation

Initial Intensity on the sample surface of exciting radiation	$I_0(\lambda)$
Factor by which this is reduced reaching layer Δx at height x	$\exp[-\mu(\lambda)\rho_m \cdot x \cdot \operatorname{cosec} \psi_1]$
Probability of excitation	$C_j \cdot \dfrac{r_j - 1}{r_j} \omega_j \cdot g_j \cdot \mu_j(\lambda)$
Fraction emitted in direction defined by collimator	$\dfrac{d\Omega}{4\pi}$
Fraction by which fluorescent radiation is reduced due to attenuation	$\exp[-\mu(\lambda_a)\rho_m \cdot x \operatorname{cosec} \psi_2]$

$I_0(\lambda)$ initial intensity on the total surface of the sample;
$\mu(\lambda)$ absorption coefficient of matrix for λ;
$\mu(\lambda_j)$ absorption coefficient of matrix for characteristic wavelength of element j;
ρ_m density of matrix;
r_j absorption jump;
x height of layer Δx above sample surface;
ψ_1 angle between mean direction of primary X-ray and sample surface;
ψ_2 take-off angle of spectrometer;
C_j weight concentration of element j;
ω_j fluorescent yield;
g_j probability of a particular electron transfer in a series.

Integration of this expression for values of x from zero to infinity gives

$$I_j = I_0(\lambda)\frac{r_j - 1}{r_j}\omega_j \cdot g_j \cdot \frac{d\Omega}{4\pi}\mu_j(\lambda)\frac{C_j \cdot \rho_m \cdot \text{cosec } \psi_1}{\mu(\lambda) \cdot \rho_m \text{ cosec } \psi_1 + \mu(\lambda_j) \cdot \rho_m \cdot \text{cosec } \psi_2}$$

$$I_j = P_j \cdot I_0(\lambda)\frac{\mu_j(\lambda) \cdot C_j}{\mu(\lambda) + A\mu(\lambda_j)} \tag{1.19}$$

where P_j is a constant for a given element and a given spectrometer and

$$A = \frac{\sin \psi_1}{\sin \psi_2}$$

This expression is only valid for completely homogeneous samples and does not allow for multiple scatter or enhancement effects. Also the integration from zero to infinity is only permitted if the sample is sufficiently thick. For the majority of samples, however, infinite thickness is rarely in excess of a few hundred microns.

1.4.3 EXCITATION BY CONTINUOUS SPECTRA

When continuous radiation is used to excite (λ_j) it is necessary to consider all of the primary wavelengths $I(\lambda)$ between the absorption edge wavelength (λ_{edge}) corresponding to (λ_j) and the minimum wavelength of the continuum (λ_{min}). Thus

$$I(\lambda) = \int_{\lambda_{\text{min}}}^{\lambda_{\text{edge}}} \cdot J(\lambda) \cdot d\lambda \tag{1.20}$$

where $J(\lambda)$ represents the X-ray tube spectrum. A general intensity formula can be obtained by substitution of this expression for $I(\lambda)$ in place of the intensity of the monochromatic radiation $I_0(\lambda)$ in Equation (1.18). It is also necessary to consider the absorption of all elements in the sample matrix on both primary and secondary wavelengths. By use of equation (1.16) the following expressions are obtained:

$$\mu(\lambda) = \Sigma C_i \cdot \mu_i(\lambda) \tag{1.21}$$
$$\mu(\lambda_i) = \Sigma C_i \cdot \mu_i(\lambda_j) \tag{1.22}$$

Substitution of Equations (1.20) to (1.22) in Equation (1.19) gives a general expression:

$$I_j = P_j \cdot C_j \int_{\lambda_{\text{min}}}^{\lambda_{\text{edge}}} J(\lambda) \cdot \mu_j(\lambda) \frac{1}{\Sigma_i C_i [\mu_i(\lambda) + A \cdot \mu_i(\lambda_j)]} \tag{1.23}$$

where $P_j = \omega_j \cdot g_j \dfrac{r_j - 1}{r_j} \cdot \dfrac{d\Omega}{4\pi}$ \hfill (1.24) and $\mu_j(\lambda)$ is the absorption by element j of the respective exciting wavelength λ.

1.4.4 SIGNIFICANCE OF THE INTENSITY FORMULA

The efficiency of producing a wavelength λ_j for a wavelength λ in the primary spectrum can be defined as

$$\frac{I(\lambda_i)}{I_0(\lambda)} = P \cdot \frac{\mu_j(\lambda) \cdot C_j}{\Sigma_i C_j [\mu_i(\lambda) + A \cdot \mu_i(\lambda_j)]} \tag{1.25}$$

For a given element and a fixed equipment geometry an efficiency factor $C(\lambda\lambda_j)$ can be introduced as

$$C(\lambda\lambda_j) = \frac{\mu_j(\lambda)}{\Sigma_i C_i [\mu_i(\lambda) + A \cdot \mu_i(\lambda_j)]} \tag{1.26}$$

The curves in Fig. (1.11) illustrate the variation in $C(\lambda\lambda_j)$ as function of the exciting wavelength λ. The first pair of curves calculated for Zn Kα are for pure zinc and for zinc as a minor (0.1 %) constituent of copper. These elements are of neighbouring atomic number and very little difference is shown between the two curves until the absorption edge is approached very closely. This can be understood by consideration of the fact that $\mu_{Cu}(\lambda_{Zn})$ has approximately the same value as $\mu_{Zn}(\lambda_{Zn})$ and further that over a large range of the values of λ, $\mu_{Zn}(\lambda)$ and $\mu_{Cu}(\lambda)$ increase in a similar manner. The second pair of curves are for Al Kα in pure aluminium and as a minor (0.1 %) constituent in a widely different matrix i.e. copper. Here the shape of the curves are very different. This can be explained by study of the absorption tables (Appendix 2b) and Equation (1.26). For a pure metal Equation (1.26) reduces to

$$C(\lambda\lambda_j) = \frac{\mu_j(\lambda)}{\mu_j(\lambda) + A \cdot \mu_j(\lambda_j)}$$

The absorption jump r has a value for aluminium of about 10 and for the majority of equipment the factor A has a value of around 1.5 to 2. This means that the second term in the denominator can be neglected and so for a pure element it follows that $C(\lambda\lambda_j)$ is almost independent of λ. However, in a heavy absorbing matrix, such as copper, the matrix itself determines the denominator, particularly the secondary absorption, and $C(\lambda\lambda_j)$ increases with $\mu_{Al}(\lambda)$. This last case is usually encountered when dealing with low concentrations in highly absorbing matrices.

Another interesting conclusion can be drawn from Equation (1.23). If the case of a heavy element in a low absorbing medium is considered then $\mu_j(\lambda) > > \mu_i(\lambda)$ and $\mu_j(\lambda_j) > > \mu_i(\lambda_j)$. In this $\Sigma_i C_i \mu_i(\lambda)$ can be neglected

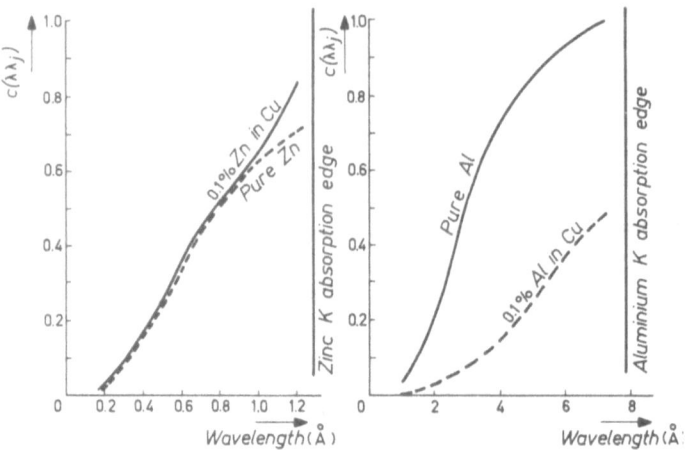

Fig. 1.11 Optimum wavelength for excitation.

compared with $C_j\mu_j(\lambda)$, provided that C_j is not too small, the resultant intensity being independent of C_j. This means, that a horizontal calibration line results and an analysis for C_j becomes impossible.

The actual intensity distribution of (λ) is composed of the continuous spectrum and the characteristic target lines. The intensity of the target lines is many times larger than that of the white continuum for any wavelength λ. The exact ratio of characteristic to continuous intensity depends on the target material and the voltage which is applied to the X-ray tube. Although its value is difficult to measure, the actual intensity distribution can be estimated by taking a reflection diagram of the spectrum scattered by a blank specimen with high scattering power, for example, a pure hydrocarbon mixture, cane sugar, graphite,[31] etc. Corrections must however, be made for the incoherent scattering. This blank diagram is also useful since it reveals any contamination of the tube spectrum and allows a check to be made of the true high voltage on the X-ray tube.

The actual fluorescent intensity is found by multiplying every value of $c(\lambda\lambda_j)$ with the appropriate value of $J(\lambda)$ and integrating over λ between λ_{min} and λ_{edge}. However $J(\lambda)$ decreases rapidly with increasing λ and the intensity diminishes in a manner given by Kramers equation (1.3), with appropriate correction for absorption by the target and the X-ray tube window. It can thus be shown that the most effective excitation is brought about by wavelengths where the product of $J(\lambda)$ and $c(\lambda\lambda_j)$ reaches a maximum. It is clear, from the foregoing that the place of this maximum, i.e. the value-of an effective wavelength, depends either on $c(\lambda \lambda_j)$ and/or on the composition of the matrix.

It is interesting to note that the relative importance of the primary and secondary absorption can be influenced by altering the value of A. For instance, A becomes very large at low take-off angles, indicating that in this case the secondary absorption effect dominates. This is an important practical consideration since secondary absorption effects are relatively easy to calculate.

REFERENCES

1. RÖNTGEN, W. C., 1898, Ann. Physik u. Chem., **64**, 1.
2. COMPTON and ALLISON, *X-rays in theory and experiment*, Van Nostrand, New York, 1935.
3. FREIDRICH, W., KNIPPING, P. and LAUE, M. VON, 1912, Ber. Bayer. Akad. Wiss., 303.
4. KULENKAMPFF, H., 1922, Ann. Physik, **69**, 594.
5. DUANE, W. and HUNT, F. L., 1915, Phys. Rev., **6**, 166.
6. KRAMERS, H. A., 1923, Phil. Mag., **46**, 836.
7. SMEATON, W. A., 1965, Chemistry in Britain, **1**, 353.
8. TSUTSUMI, K., 1959, J. Phys. Soc. Japan, **14**, 1696.
9. KAKUSCHADSE, T. J., 1959, Ann. Physik., **3**, 352.
10. AUGER, P., Thesis, Paris, 1926.
11. BURHOP, *The Auger Effect*, University Press, Cambridge, 1952.
12. KUHN, *Atomic Spectra*, Longmans, London, 1962.
13. JOPSON, R. C., MARK, H., SWIFT, C. D. and WILLIAMSON, M. A., 1963, Phys. Rev., **131**, 1165.
14. GLOCKER, R. and SCHRIEBER, H., 1928, Ann. Physik., **85**, 1089.
15. FOX, J. G. M., 1963, J. Inst. Metals, **91**, 239.
16. HANS, A., HANCART, J. and HOUBART, I., *Analyse par les rayonnements X*, (Bruxelles, 1964), Philips, Eindhoven.
17. BIRKS, *Electron Probe Microanalysis*, Wiley, New York, 1963.
18. CAMERON, J. F. and RHODES, J. R., 1961, Nucleonics, **19**, 53.
19. SEIBEL, G., TRAAON, J. Y. and MARTINELLI, P., 1961, Rev. Universelle des Mines, **18**, 260.
20. WATT, J. S., 1964, The International Journal of Applied Radiation and Isotopes, **15**, 617.
21. HENKE, *Advances in X-ray analysis*, Plenum, New York, 1961, **5**, 288.
22. WYKOFF, R. W. G. and DAVIDSON, F. D., 1964, Rev. Sci. Instr., **35**, 381.
23. DUNNE and MILLER, *Developments in Applied Spectroscopy*, Plenum, New York, 1964, **4**, 33.
24. BERNSTEIN, *Developments in Applied Spectroscopy*, Plenum, New York, 1964, **4**, 45.
25. VICTOREEN, J. A., 1949, J. Appl. Phys., **20**, 1141.
26. *International Tables for X-ray Crystallography*, Kynoch, Birmingham, 1962, **3**.
27. KLEIN, O. and NISHINA, Y., 1928, Z. Physik, **52**, 853.
28. COMPTON, A. H., 1923, Phys. Rev., **21**, 485.
29. BIRKS, L. S., 1960, Spectrochim. Acta, 148.
30. SPIELBERG, N., 1959, Philips Research Reports, **14**, 215.
31. MÜLLER, R., 1962, Spectrochim. Acta, **18**, 123 and 1515.

CHAPTER 2

DISPERSION

2.1 General

The basic function of the spectrometer is to provide a means of isolating a selected wavelength from the polychromatic beam of characteristic radiation excited in the sample, in order that individual intensity measurements can be made. Although this is normally achieved by making use of the specific diffracting property of large single crystals, this is not by any means the only way of selecting a specific wavelength range and other methods which have been employed include the use of diffraction gratings,[1] balanced filters[2-3] and energy resolution in the form of pulse height selection. The usual wavelength range of the conventional X-ray spectrometer is between 0.2 to 15 Å and over this region the single crystal is certainly the most efficient and versatile means of dispersion, particularly in combination with pulse height selection for the removal of harmonic overlap (See Chapter 4). However, the recent successful attempts to extend the operating range of the X-ray spectrometer into the soft X-ray and vacuum ultra-violet region have provided greater incentive for a more detailed study of the use of gratings in this region as well as the exclusive use of pulse height selection. Since measurements in the soft X-ray region invariably require some modification to the commercially available spectrometer, usually by way of modification to the source of primary radiation and the detector, it is the intention to first consider dispersion in the conventional wavelength range and then to discuss the soft X-ray region as a separate topic.

2.2 Geometric arrangement of the spectrometer

The theory of X-ray diffraction has already been treated in Chapter 1, and what must now be considered is the way in which this property can be variously employed in order to achieve the optimum requirement of the spectrometer i.e. wavelength resolution without excessive intensity loss. A spectrometer consists basically of a system of slits or collimators and a monochromator, in this case the analysing crystal. Numerous geometric arrangements have been employed in X-ray spectrometer and Table 2.1 summarises these various arrangements along with their more important advantages and disadvantages.

The two most important categories are illustrated in Fig. 2.1, these being the flat crystal arrangement (a) and the curved crystal arrangement (b). In the flat crystal system[4] a portion of the secondary radiation is selected by a primary collimator and the parallel beam is allowed to fall into the plane surface of a single crystal which has been cleaved in such a way that diffracting planes form a tangent to the focussing circle. Here the radiation is diffracted in accordance with the Bragg relationship (see Section 1.3.5) i.e. $n\lambda = 2d \sin \theta$ where λ is the wavelength of the radiation diffracted through an angle θ, by planes in the analysing crystal of spacing d; n is an integer. After diffraction the radiation passes through a secondary collimator to the detector.

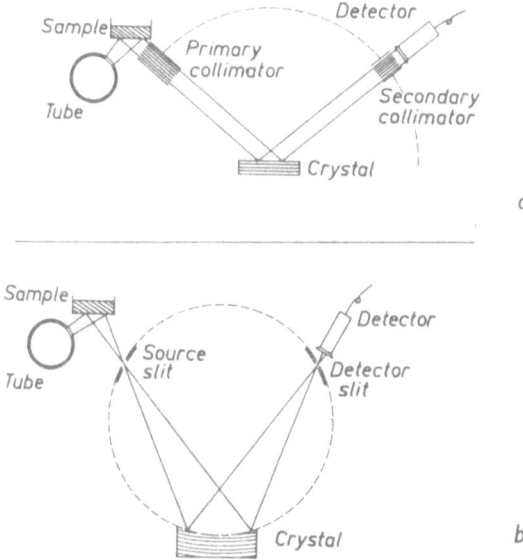

Fig. 2.1a Plane crystal spectrometer geometry.
Fig. 2.1b Curved crystal spectrometer geometry.

The Bragg condition will be satisfied as long as the relative angular rotation of the detector is kept to twice that of the analysing crystal. The fundamental disadvantage of this system is that the primary collimation necessarily rejects a very high proportion of the incident radiation and intensity losses are high.

TABLE 2.1

Geometric arrangements of X-ray spectrometers

System	Optics	Advantages	Disadvantages
Flat Crystal[4]	Crystal forms tangent to focussing circle.	Simple geometry, line broadening not too dependent upon θ.	Collimation necessary. Resolution limited by divergence allowed by collimator and rocking angle of crystal.
Johann[5]	Crystal bent to radius R.	X-rays diffracted to line focus, therefore only a simple scatter slit is required. Fairly large acceptance angle.	Defocussing occurs with $\Delta\theta$. Since $U = V = R \sin \theta$, as θ changes either V must change (i.e. an irregular detector path is necessary) or a range of R values must be employed.
Johansson[6]	Crystal bent to radius R, ground to radius $R/2$.	Similar to Johann but not defocussing.	Similar to Johann except for defocussing due to crystal-focussing circle divergence.
Variable curve[7]	Crystal bent automatically such that $R \propto \sin \theta$.	Only one crystal required for large wavelength range.	Some defocussing is inevitable. Only a limited range of crystals are available which can be elastically deformed.
Cauchois[8]	Crystal bent to radius R. Used in transmission. Planes normal to surface diffract.	Large solid angle of acceptance.	Great attenuation of X-ray beam as it passes through the crystal.
Crystal Edge[9]	Crystal planes immediately parallel to crystal edge diffract whole spectrum at once.	Only detector moves therefore simple to construct. Film record of whole spectrum can be taken at same time.	Intensities are low. Different wavelengths arise from different areas of the specimen.

In the curved crystal arrangement (Johann condition)[5] the analysing crystal is bent to a radius R and lies on the edge of the focussing circle of radius $R/2$. Radiation arriving at the crystal from a distance greater than $R/2$ will pass through a point of focus on the focussing circle and will be diffracted through a similar point of focus at a distance equal and opposite to that of the incident radiation. In practice as the incident radiation arises from an area on the specimen surface the foci will be lines rather than points. A disadvantage of this system is that defocussing occurs due to the finite length of the crystal, since towards the extreme edges of the crystal the diffracting planes are not tangential to the focussing circle. This defocussing can be avoided by bending the crystal to radius R and grinding to radius $R/2$ as in the Johannson[6] condition, or minimised by automatically bending the crystal such that R is proportional to sin θ.[7] A fundamental disadvantage with the curved crystal system is that spherical aberration occurs unless the condition $U = V = R \sin \theta$ is maintained and the advantages which have been gained over the flat crystal system by doing away with collimators are lost by the need to use a detector slit to maintain line profile shape. A practical disadvantage is that either the curvature of the crystal surface or its distance from the specimen has to be varied with wavelength, hence in practice the use of curved crystal optics is generally restricted to spectrometers of fixed geometry. Since the flat crystal system is the most widely employed in commercially available spectrometers which have been designed for general routine use, it is the intention to limit discussion specifically to this system although many of the factors which will be discussed apply equally to other geometric arrangements. For a more comprehensive comparison of the relative merits of the various optical systems employed the reader is recommended to consult other works covering this field.[10-12]

2.3 Effective range of spectrometer

In order to successfuly disperse radiation in the 0.2 to 15 Å range it is necessary that suitable d value analysing crystals be available, with due consideration to the angular range over which the spectrometer will function. The upper limit of the spectrometer is normally fixed by the angle at which further movement of the detector system is prevented by the presence of the sample chamber. The magnitude of this particular angle will depend upon the particular spectrometer design but in general, is around 145° 2 θ. The lower limit is set by the angle at which the analysing crystal no longer intercepts the majority of the incident radiation. Fig. 2.2 demonstrates that this angle is dependent upon the length L of the analysing crystal and the width b of the incident beam. Typical values for b and L are 12.5 mm and 65 mm respectively giving a θ value of $\sin^{-1}\left[\dfrac{12.5}{65}\right] = 11.1°$ or $22.2°\ 2\ \theta$. Where crystals are to be used exclusively for dispersion of long wavelength radiation it is common practice to use an L value of about 40 mm and in this instance the minimum $2\ \theta$ angle is equal to $36.4°$. It will be apparent from the Bragg relationship that the maximum wavelength which can be diffracted by a certain crystal plane is given by $(\lambda)\,\text{max} = 2d \cdot \sin 72.50° = 1.91d$. In addition any wavelength limitation occurs at the long wavelength end of the spectrum since shorter wavelengths can still be diffracted within the normal range of the spectrometer.

Table 2.2 lists the more commonly used analysing crystals along with their useful ranges and average reflection efficiencies. In general, an analysing crystal is selected on the basis of both optimum dispersion and reflection efficiency. Since both the dispersion and measured intensity are also functions of collimation it is necessary to consider the synergistic effect of collimators and crystal when selecting optimum dispersion conditions (c.f. Appendix 2a).

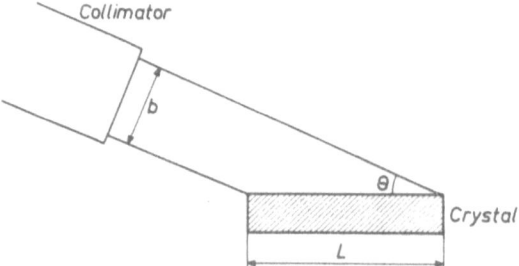

Fig. 2.2 Optimum length of crystal analysers. A beam of radiation of width b falls onto the analysing crystal of length L, making an angle θ, where θ is equal to $\sin^{-1}(b/L)$. When the goniometer is set at an angle below this value the crystal can no longer intercept the whole of the incident beam.

2.4 Dispersion efficiency

The dispersion efficiency of a spectrometer is fixed by the spacing of the dispersion medium i.e. the d spacing of the crystal, the line width of the measured profile and the solid angle of acceptance of the detector system. The dispersion of the analysing crystal can be derived by differentiating Equation 1.17, i.e.

TABLE 2.2

Analysing Crystals

Crystal	Reflection Plane	2d Spacing (Å)	Lowest Atomic Number Detectable		Reflection Efficiency
			K Series	L Series	
Topaz	(303)	2.712	V (23)	Ce (58)	Average
Lithium Fluoride	(220)	2.848	V (23)	Ce (58)	High
Lithium Fluoride	(200)	4.028	K (19)	In (49)	Intense
Sodium Chloride	(200)	5.639	S (16)	Ru (44)	High
Quartz	(10$\bar{1}$1)	6.686	P (15)	Zr (40)	High
Quartz	(10$\bar{1}$0)	8.50	Si (14)	Rb (37)	Average
Penta erythritol	(002)	8.742	Al (13)	Rb (37)	High
Ethylenediamine Tartrate	(020)	8.808	Al (13)	Br (35)	Average
Ammonium Dihydrogen Phosphate	(110)	10.65	Mg (12)	As (23)	Low
Gypsum	(020)	15.19	Na (11)	Cu (29)	Average
Mica	(002)	19.8	F (9)	Fe (26)	Low
Potassium Hydrogen Phthalate	(10$\bar{1}$1)	26.4	O (8)	V (23)	Average
Lead Stearate		100	B (5)	Ca (20)	Average

$$\frac{d\theta}{d\lambda} = \frac{n}{2d} \times \frac{1}{\cos\theta} \qquad (2.1)$$

From this it will be seen that dispersing power increases with increase of θ or decrease of d and in addition, the higher the order of the reflection the better the dispersion. It is also apparent that resolution problems will be particularly prevalent at low angles, especially since due to the fact that λ varies as $\frac{1}{Z^2}$ the relative wavelength separation of neighbouring atomic numbers decreases with decrease of θ for a fixed d value crystal. The width of the measured line profile depends upon the natural line width, the broadening due to the analysing crystal and the degree of primary and secondary collimation.

2.5 Broadening of line profile

The X-ray K-emission line does not present one specific wavelength, but rather may be contained in a narrow band of wavelengths with a bandwidth of the order of some tenthousands of an Ångström. This natural line width is appreciably larger for light elements and may result in an asymmetric maximum as, for example, in the case of magnesium. A large single crystal may be considered to consist of many small blocks or regions, each one having the same crystal structure but being slightly misaligned with respect to the others. If perfect alignment existed throughout the whole analysing crystal the total reflected intensity would be much less than in the case of this so-called mosaic structure. Each block may, in itself, show a small difference in diffracting angle θ, compared with another block under the same conditions, due to small differences in structure or internal strain. The total analysing crystal will thus not only reflect the monochromatic X-rays at an angle 2 θ, but rather over a region 2 $\theta \pm \alpha$. The curve giving the relation between reflected intensity and angle 2 θ as defined by the Bragg equation, approximates to Gaussian distribution. A good analysing crystal should have a narrow rocking curve but a high integrated reflected intensity. In practice, most analysing crystals show a curve with a width at half maximum intensity of 0.2-0.4° (2 θ). The orientation and properties of the blocks determine the height and the shape of the curve and so the reflecting power of a crystal may thus be influenced by surface treatment.

2.6 Collimation requirements of the spectrometer [4,13]

For a spectrometer of given geometry, increase in dispersion efficiency can only be obtained at the expense of loss of intensity and consequently some compromise between these two factors must be reached in the design of a spectrometer. Invariably the requirement is for high intensity at moderate peak to background and most modern spectrometers are designed on this basis. However, the resolution requirements also depend to a certain extent on the potential overlap of lines which cannot be rejected by pulse height selection, in general, first and second order lines. Since the relative wavelength dispersion between neighbouring atomic numbers increases with decrease of atomic number, high collimation is not so important for the lower atomic number elements. As fluorescent yield values are low and absorption losses high, there is a great need to use the minimum of collimation in the long wavelength range. To this end manufacturers frequently supply two interchangeable primary collimators, such that the degree of collimation can be chosen to suit the requirements for the particular wavelength range being measured (c.f. Appendix 2a).

The most widely applied collimator systems are based on the parallel plate method first described by Soller.[14] Since the purpose of the collimation system is to limit the divergence of the X-ray beam and provide acceptable angular resolution, the distance between the plates of a given collimator must be calculated such that the theoretical collimator divergence embraces the width of the diffraction profile of the crystal. In general, this entails a spacing between plates of 200-500 μ for a primary collimator.

The selection of the optimum collimator for a certain application must be made with due consideration to the dispersion of the analysing crystal employed.[15] A choice is frequently available between, on the one hand, a fine collimator and a crystal of poor dispersion but high reflectivity and on the other hand, a coarse collimator and a crystal of high dispersion but poor reflectivity. Fig. 2.3 illustrates this with reference to the resolution of manganese Kα and chromium Kβ. Fig. (2.3D) shows that the lines are insufficiently resolved when using a lithium fluoride (200) crystal and a 160 μm collimator. Use of a topaz crystal (Fig. 2.3B) completely resolves the lines but the overall intensity drops by a factor of about 5. However, by use of a 480 μm collimator (Fig. 2.3A) this factor is reduced to 1.8 and lines are sufficiently resolved for measurement.

The need for a high degree of secondary collimation is normally unnecessary although, as will be discussed later, sufficient collimation must be applied to prevent spurious reflections from the crystal from reaching the detector. Very high resolution can, however, be obtained by application of secondary collimation by virtue of decreasing the solid angle of acceptance of the detector. For example, it is possible to completely resolve chromium Kβ and manganese Kα diffracted with a lithium fluoride crystal by use of a 10 cm long secondary collimator with 150 μm spacings. However, the overall intensity is decreased by a factor of about 4 and in addition, very high angular reproducibility of the goniometer is required if this degree of collimation is to be employed.[16]

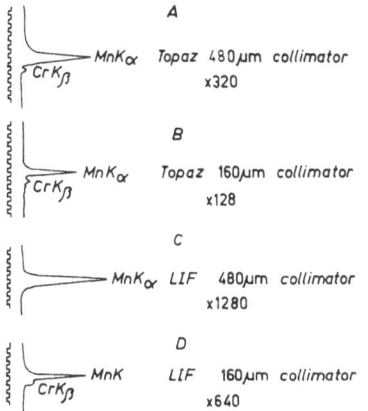

Fig. 2.3 Resolution as a function of crystal dispersion and collimation.
X-ray spectrography chart of a metal powder specimen, recorded with a fine and a coarse collimator, using two different analyzing crystals.
The charts show the difference in angular resolution and in intensity of the various combinations.

A = Topaz   crystal with 0.48 mm collimator
B = Topaz   crystal with 0.16 mm collimator
C = LiF(200) crystal with 0.48 mm collimator
D = LiF(200) crystal with 0.16 mm collimator

2.7 Reflection efficiency

It has been mentioned already that a perfect crystal invariably has a poor reflection efficiency. An "ideally imperfect" crystal is a far better diffracting agent and is considered to be made up of thousands of small crystal fragments in completely random arrangement. If the fragments are small enough each one will remove a very small amount of the incident energy and since no phase relationship exists between the waves diffracted from different fragments the total reflection will be made up from the sum of all the reflections from the small fragments.[17] In the preparation of analysing crystals the surfaces are often deliberately abraded in order to optimise between the best reflection efficiency and a suitable rocking curve. This surface abrasion can be applied in a variety of ways ranging from rubbing the surface of the crystal with coarse emery paper to careful ultrasonic treatment.

The reflection efficiency of an analysing crystal is often dependent upon the wavelength of the radiation which it is diffracting and this is due to a number of reasons. One of the more important of these depends upon the fact that diffraction occurs by a process of scattering of the X-rays by the individual atoms making up the crystal. The scattering power of one individual atom, f, depends on the number of electrons combined in the orbitals of this atom, the angle of diffraction θ and the wavelength of the primary X-rays. The waves scattered by the individual atoms will combine to give rays reflected in directions given by Bragg's law, as this equation selects sets of lattice planes, indicated by three numbers $(h\ k\ l)$ and the angular positions of this lattice planes for reflection. However, the intensity of these reflected rays, depends upon the scattering power f, of the individual atoms and the phase differences between the individual waves. This phase difference is given by the crystal structure, by the arrangement of the individual atoms, and by the orientation of the reflecting lattice planes in the unit cell. This means that the reflection efficiency of an analysing crystal depends on the crystal structure, the kind of atoms, the reflecting planes, the order of reflection and the wavelength of the X-rays to be reflected. The reflection efficiency of a given analyzing crystal may vary slightly with increasing order of reflection, as with LiF, or may show sudden changes as with Topaz, depending on the crystal structure.

The wavelength dependence of the reflecting power may vary considerably from one analyzing crystal to another (although much of the effect is often due directly to the length of the analyzing crystal of fig. 2.3). For instance, the reflectivity of topaz (303) crystal compared to that of LiF (200) is

20% for copper radiation (1.54 Å), 27% for molybdenum (0.71 Å) and 39% for barium (0.39 Å).

The implications of this in X-ray spectrometry are obvious, where gain in intensity must be balanced against loss in resolution. For instance, in the region of BaKα, the gain in resolving power outweighs the loss in intensity (only 2.5 times compared with LiF (200)), and the topaz crystal will generally be preferred. For copper radiation, however, the intensity gain dominates and LiF (200) will be given preference. Another LiF crystal, cut with the (110) planes parallel to the crystal surface, has resolving power similar to that of topaz and a reflectivity which is much higher.

A consequence of the dependence of the scattered intensity on the crystal structure is the possibility to select certain analyzing crystals, cleaved such that the second order is missing. For example, a crystal of silicon, cleaved in such a way that the (111) planes ($2d = 6.27$ Å) are parallel to the crystal surface, gives practically no second order reflection at all. Its third order reflection is in turn 10% the intensity of the first. These crystals sometimes find useful application where second order overlap cannot be removed by conventional means.[18]

2.8 The use of filters to increase resolving power

The absorption of X-radiation by a metal sheet is strongly dependent upon the thickness of the sheet and its absorption coefficient for X-rays of the particular wavelength (c.f. Equation (1.9); Fig. 1.7 and Appendix 1).
The absorption coefficient of a pure element differs by a factor of 4-6 between the short and long wavelength sides of the K-discontinuity (the K-edge). X-radiation just to the short wavelength side of the edge will be strongly absorbed, whereas the absorption of X-radiation just to the long wavelength side is much less. Thus in principle it is possible to use this property to eliminate unwanted radiation.

2.8.1 REDUCING SECONDARY RADIATION

If a suitable filter is placed somewhere in the optical path between specimen and detector, radiation emitted or reflected by the sample can be almost completely eliminated. In this way an analysis line can often be freed from interference arising from somewhat harder radiation. Very often the most convenient position for the filter is immediately in front of the detector. An example might be the analysis of small quantities of tantalum in the presence of very high concentrations of tungsten. Under normal conditions the broad maxima of WLα at 1.48 Å will partially mask the small peak of TaLα at 1.52 Å. The absorption edge of nickel is situated at 1.49 Å and by use of a thin (20 μ) sheet of nickel the tungsten radiation is almost completely absorbed whereas the tantalum intensity is only halved. The result is a clearly resolved tantalum peak whose intensity can easily be measured.

2.8.2 REDUCING PRIMARY RADIATION

Very often a characteristic X-ray line arising from the tube and reflected by the sample, interferes with a line emitted by the sample. The intensity of this characteristic tube line can usually be reduced by placing a filter directly over the X-ray tube window. However, this filter will also absorb primary X-rays from the white continuum and thus the excitation efficiency for the element to be analysed will be much less. The thickness of the filter has to be chosen in such a way that the tube line is reduced as much as possible, but still allowing sufficient harder photons to reach the sample to excite the atoms of the element to be analysed. The use of a titanium filter over the window of a chromium anode X-ray tube for the analysis of small quantities of chromium is an example of the use of this type of filter (c.f. Section 9.12).

2.9 Problems experienced in the application of crystal dispersion

It is important that the spectroscopist be aware of potential pitfalls in the application of crystal dispersion in X-ray spectrometry. The success with which a particular crystal can be employed depends not only upon its crystallographic suitability but also to a large extent on its physical properties. For example, whereas the appearance of spurious lines is due exclusively to the crystallographic properties of the crystal, the variation in reflection efficiency with wavelength is essentially a physical property, as is the variation of d spacing with temperature. All of these factors have to be considered in the selection of new crystals as well as in the application of existing ones and the following is a short resume of the more important effects.

2.9.1 GENERAL CONDITION OF THE CRYSTAL

It is important that the general state of the crystal should be good both from the point of view of its physical and crystallographic condition. The crystal should be free from uneven strains and the effect of any surface treatment should be equivalent over the whole surface area. This is particularly important since the irradiated area of the crystal surface varies with the angular setting of the spectrometer. It is also important that the crystals be free from twinning since this in turn will lead to doubling-up of the diffracted peaks. Crystals such as sodium chloride which are hygroscopic should be kept in a desiccator when not in use, and crystals such as gypsum, which tend to effloresce should not be kept under high vacuum conditions for unnecessarily long periods. Since analysing crystals are invariably expensive there is an additional incentive in insuring that they are carefully looked after, and it is good practice to store unused crystals in a desiccator. Certain crystals such as ethylene diamine d-tartrate and penta-erythritol tend to deteriorate with time particularly if subjected to high radiation fluxes. From this point of view it is good practice to periodically check the reflection efficiency of all crystals which are in regular use.

2.9.2 TEMPERATURE EFFECTS

The purpose of the goniometer is to accurately reproduce the angle at which each wavelength occurs. The accuracy with which this can be fulfilled will depend not only upon the mechanical construction of the goniometer but also upon the magnitude of any variations in the d spacing of the analysing crystal. It is important, therefore, that the thermal expansion coefficient of each crystal be established with respect to the d spacing employed. These data have been measured in our laboratories for most of the commonly used crystals and Fig. 2.4 shows the correlation between change in angle per degree centigrade, with angle. As would be expected the magnitude of the effect of the expansion coefficient on the angular shift increases with diffraction angle. Fortunately this effect is counteracted to some extent by the fact that the width of the diffracted line also increases with increase of diffraction angle, hence angular reproducibility is less critical at high goniometer angles. In general, for the conventional X-ray spectrometer, an angular reproducibility of about $0.01°$ 2θ is sufficient up to about $60°$ 2θ and around $0.02°$ 2θ between $60°$ and $150°$ 2θ. Assuming a total temperature variation of less than five centigrade degrees all of the crystals examined have the necessary temperature stability with the exception of penta-erythritol. This crystal is particularly temperature sensitive and the effect on angular reproducibility becomes very significant above $100°$ 2θ. In effect this means that when using penta-erythritol to disperse aluminium $K\alpha$ radiation ($2\theta \simeq 145$ °C), and to a lesser extent silicon $K\alpha$ radiation ($2\theta \simeq 109°$) special precautions must be

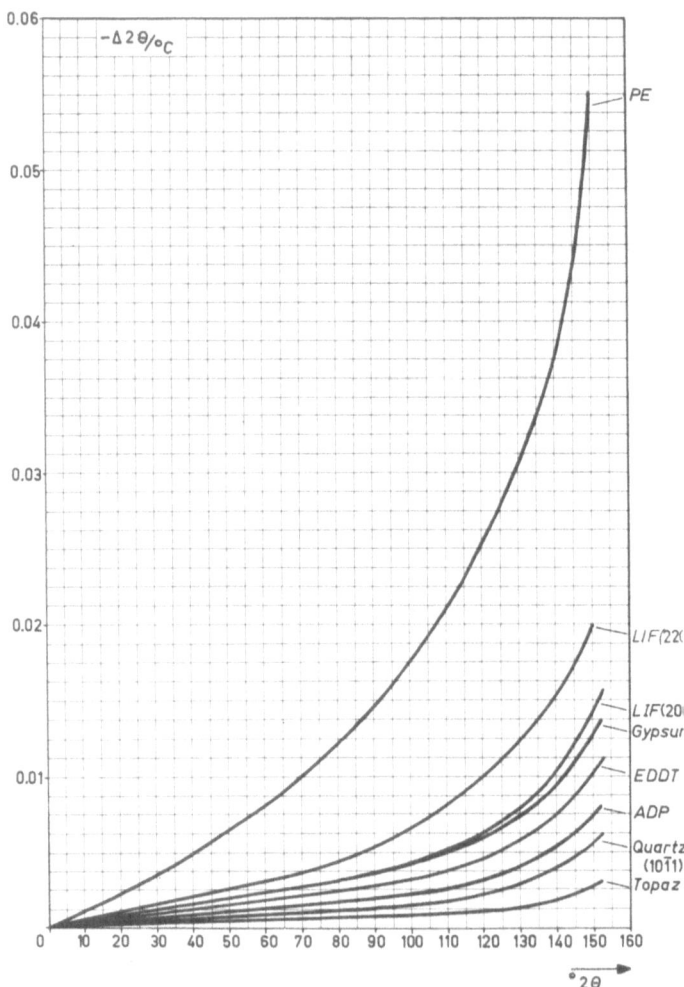

Fig. 2.4 Temperature sensitivities of crystal analysers. The curves show the correlations between the change in 2 θ value per degree centigrade and angular setting, for most of the commonly used crystal analysers.

taken to minimise the angular shift with temperature. This invariably entails the frequent re-peaking of the crystal since close crystal temperature control is difficult owing to the large thermal inertia of the spectrometer.

2.9.3 CRYSTAL FLUORESCENCE

As the analysing crystal necessarily lies directly in the path of the radiation leaving the primary collimator, all of the atoms making up the crystal matrix are being bombarded with a fairly high radiation flux and are themselves sources of fluorescent radiation. If this radiation is sufficiently energetic to reach the detector, an increased background count rate will result over the whole of the angular range of the spectrometer. The magnitude of this increased background will be dependent only upon the scattering power of the sample, the efficiency of the detector and the attenuation by the path length between crystal and detector. In general, this effect is not significant when measuring radiations of less than about 3.5 Å [e.g. of atomic number greater than 20 (calcium) when measuring K radiations] because the elements making up the more commonly used crystals for use over this range are of sufficiently low atomic number. The problem does become important, however, in the measurement of softer radiations and of the crystals normally used, sodium chloride, potassium hydrogen phthalate, ammonium dihydrogen phosphate and gypsum (calcium sulphate) all give rise to significant crystal fluorescence. Since the fluorescence radiation from the crystal is essentially monochromatic, consisting in the main of $K\alpha$ and $K\beta$ radiations, it is often possible to remove it by pulse height selection provided that the energy difference between the measured wavelength and the wavelength of the crystal fluorescence is sufficiently large. This particular procedure is described in detail in Chapter 4.

2.9.4 ABNORMAL REFLECTIONS

The occurence of extra reflections, i.e. reflections not directly or apparently attributable to the sample, can sometimes cause difficulty in the interpretation of X-ray spectra. These extra reflections may arise from coherently or incoherently scattered lines from the X-ray tube anode, from counter tube effects[19] or from abnormal reflections from the crystal.[20] The magnitude of tube lines and counter tube effects are easily assessed by running a "blank" spectrum of, for instance, distilled water. Abnormal reflections, however, are difficult to assess, since they are due to lines from the sample being reflected at angles other than those predicted from the d spacing of the analysing crystal.

This is not a common effect and is in general restricted to analysing crystals of fairly low symmetry such as topaz [21] (orthorhombic) and ethylenediamine d-tartrate (monoclinic). Circumstances can arise, however, whereby extra reflections do reach the detector, particularly when the degree of secondary collimation is insufficient. It is necessary that the origin of these spurious reflections be understood in order that steps can be taken to avoid or minimize them.

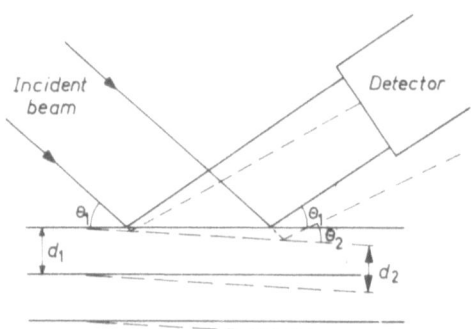

Fig. 2.5 Origin of extra crystal re-flections.

Fig. 2.5 illustrates one reason for the occurrence of extra reflections and shows an incident beam striking the plane surface of the crystal at angle θ_1, and being diffracted in the normal way by the planes of spacing d_1 to the detector. In the circumstances only wavelengths satisfying the Bragg condition for d_1 and θ_1 will enter the detector. A second set of crystal planes of spacing d_2 and slightly misoriented with respect to planes d_1, also diffract radiation and because of the close proximity of the detector and the lack of secondary collimation, part of this diffracted beam leaving the planes d_2 at angle θ_2 also enter the detector. In this instance the radiation has a wavelength satisfying the Bragg condition for d_2 and θ_2. The overall effect is that extra radiation enters the detector having a wavelength different from that predicted by the angular setting of the goniometer and the d spacing of the crystal.

Fig. 2.6 shows part of a spectrum diffracted with topaz where abnormal reflections occur in the region of the lead L lines. These abnormal reflections are almost certainly due to diffraction of tin $K\alpha$ and tin $K\beta$ from another set of planes in the crystal which are not parallel to the crystal surface. The abnormal reflections are very much broader than normal reflections and this offers a simple method of identification. In this particular case the spectrum was recorded using a tandem type detector consisting of an argon-methane flow counter and a scintillation counter. It is common practice to also have an extra collimator between the two detectors in a tandem system of this type. By measuring the output of the scintillation counter alone, the spurious lines were completely rejected owing to the combined effect of the collimator and the extra distance of the scintillation counter from the crystal compared to the flow counter.

As far as topaz is concerned, abnormal reflections arise from slightly misoriented planes along the maximum length of the crystal which is cleaved to the (303) plane. Ethylene diamine d-tartrate shows similar effects due to

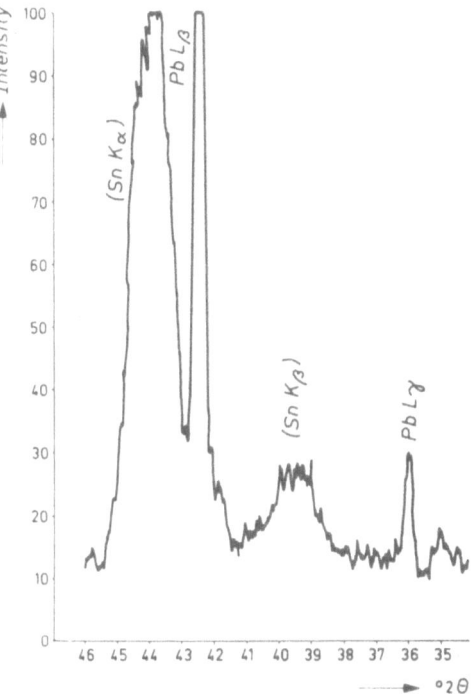

Fig. 2.6 Spurious reflections from topaz. The spectrum was recorded from a Gun-Metal sample using a topaz crystal and a proportional counter very close to the crystal surface. In addition to the normal PbL_β and PbL_α reflections, two very broad peaks are seen which are spurious reflections of SnK_α and SnK_β.

misoriented planes across the breadth of the crystal cleaved to the (020) plane and in this case extra collimation is of little help. In both cases, careful application of pulse height selection may minimize the effect and in certain instances rotating the crystal through 180° may completely eliminate the effect.

A further example of the effect of abnormal reflections is that experienced in the analysis of specimens consisting of single crystals.[22] Fig. 2.7 shows a portion of the spectra obtained[23] from samples of rutile. The lower spectrum is of poly-crystalline rutile and is devoid of peaks except for a broad maximum at about 16° 2 θ. The upper spectrum (1) is of a single rutile crystal and shows many peaks arising from Laue type diffraction of the incident beam by the sample. Spectrum (2) shows the effect of rotating the rutile crystal through a few degrees of arc about an axis perpendicular to the irradiated face of the sample and demonstrates that the intensity of the spurious reflections is a function of sample position. Great care must be taken in the interpretation of data from single crystals and several spectra should be recorded at different orientations of the sample in order to differentiate between Laue type reflections from the sample and true Bragg reflections. Wherever possible samples should be rotated during analysis since this helps to even out some of the extraneous effects. The application of pulse height selection also helps in reducing the effect of spurious lines since these frequently have energies far in excess of the true wavelength corresponding to a particular goniometer setting.

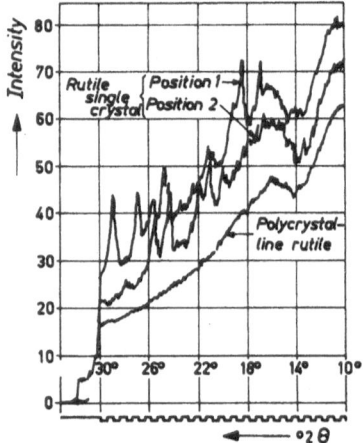

Fig. 2.7 Comparison of backgrounds for single crystal and polycrystalline samples.

2.10 Dispersion of soft X-rays

The soft X-ray region is usually considered to be the range of wavelengths from about 15 Å to the beginning of the vacuum ultra violet region (about 200 Å). Dispersion of X-radiation in this range by conventional organic or inorganic crystals is not possible since it is difficult to find suitably large d spacing crystals which give good diffraction efficiency combined with stability and freedom from unwanted reflection. As a result of this, special techniques have been developed for the dispersion of soft X-radiation and these fit conveniently into three categories, which employ respectively organo-metallic compounds, pseudo crystals and diffracting gratings.

2.10.1 ORGANO-METALLIC COMPOUNDS

A great portion of stable large unit cell crystals are organo-metallic compounds and several systematic searches have been made in an attempt to find crystals suitable for X-ray dispersion.[24-25] The most promising results todate have been obtained from the salts of phthalic acid including those of K+, Rb+, Cs+, Tl+, Na+ and NH₄+. The most useful of these crystals is potassium hydrogen phthalate (KAP) the (10$\bar{1}$1) plane of which is a cleavage plane with $2d = 26.4$ Å, this allows the dispersion of Kα radiations down to atomic number 8 (oxygen). Its reflection efficiency is good, being comparable to that of ethylenediamine d-tartrate but like all organo-metallic crystals it exhibits significant crystal fluorescence and care must be exercised in its general application to the dispersion of wavelengths at or near the wavelength of potassium K radiations.

2.10.2 PSEUDO CRYSTALS

Several techniques have been described in which pseudo crystals have been fabricated where the final d spacing is artificially contrived by selection of individual layers of suitable size. One of these methods involved the building of multi-layer sandwiches of two different metals by successive vacuum deposition. Dinklage and Frericks[26] describe the use of such a device consisting of 140 alternate layers of lead and magnesium giving a $2d$ value of 54 Å. They claim sharp reflections with a diffraction efficiency for magnesium Kα equivalent to around 5% that of ammonium dihydrogen phosphate. Although this technique offers a potentially powerful method of producing pseudo crystals of almost any d spacing it suffers from the fundamental disadvantage that inter-diffusion of the two layers can cause a rapid decrease in efficiency. It is possible that this could be minimised by careful choice of layer metals.

A more generally used method of producing pseudo crystals depends upon the building up of successive mono-layers of metal stearates using the Langmuir-Blodgett dipping method.[27] This method depends upon the fact that when an organic solution of stearic acid is added to water containing metal ions, molecules of stearate salt are produced at the surface of the water which are oriented such that the hydrophilic end of the molecules dip into the water and the hydrophobic ends are uppermost. When a hydrophobic plate is dipped into the solution, individual molecules attach themselves to the plate with their hydrophilic ends furthest from the surface of the plate. On withdrawing the plate a second set of molecules attach themselves to the mono-layer already deposited but in the reverse direction. By successively dipping and withdrawing the plate a series of mono-layers can be built up having a d spacing twice that of the length of the stearate molecules. As the stearate chain is roughly 25 Å long the distance between successive rows of metal atoms is about 50 Å giving a pseudo crystal with $2d$ of approximately 100 Å. Mica is a very convenient substate for these layers and it is easily rendered hydrophobic by rubbing with ferric stearate. It has been found that 50 to 100 mono-layers gives a sufficient depth of pseudo crystal and this number is fairly easy to achieve provided that care is taken to ensure correct conditions of pH and freedom from dust particles. Metal salts of C_6-C_{10} saturated fatty acids are also reported to show promise as potential sources of pseudo crystals,[28] the quoted $2d$ values ranging between 30-60 Å.

2.10.3 DIFFRACTION GRATINGS

It is not surprising that in the attempts to extend the usuable X-ray region into the vacuum ultra violet, attempts have been made to apply dispersion devices which are generally applied in the U.V. region. Franks[29] has recently discussed the feasibility of the use of ruled gratings in the 20-200 Å region and in a later paper Sayce and Franks[30] describe methods for producing suitable gratings based on both etching and replicating, and photographic procedures. The relationship between λ and diffracted image is given by

$$n\lambda = 2d \sin \frac{2\psi - \omega}{2} \cdot \sin \frac{\omega}{2} \qquad (2.2)$$

where d is the grating spacing, ψ the angle between the incident beam and the grating and ($\psi + \omega$) the angle between the grating and the diffracted beam. Since it is necessary that ψ should not appreciably exceed the critical angle of total reflection, values of around 1° are normally necessary for glancing angle of incidence. With grating spacings of around 2×10^4 lines per cm, the angles which would correspond to the usual 2 θ angle setting of a crystal are of the order of 2°-5° for a diffraction grating. The minuteness of these diffraction angles can lead to wavelength dependent intensity variations due to small imperfections in the grating surface and this combined with the poor diffraction efficiency resulting from the low total energy collection of a grating, have tended to limit their usefulness generally in long wavelength X-ray spectrometry.

2.11 Comparison of the methods for long wavelength dispersion

The question naturally arises as to which of the methods available for the dispersion of long wavelength radiation is the more applicable to conventional X-ray spectrometry. Purely from the point of view of convenience there is little doubt that the X-ray spectroscopist would prefer a system which would allow him to rapidly and reversibly convert his conventional spectrometer for long wavelength measurements and from this aspect, conventional or pseudo crystal is to be preferred to the diffraction grating. In addition to this, from the limited data available, it is apparent that the count rates obtainable with the best gratings are at least an order of magnitude lower than those obtainable with the best stearate crystals. At this point in time, the choice appears to lie mainly between the metal stearate crystal, the metal phthalate crystal or, where applicable, the conventional organic or inorganic crystal. Table 2.3 compares the count rate data obtained for aluminium, magnesium, sodium, fluorine and potassium $K\alpha$ radiations, using the various types of crystal which are available. It will be seen that whereas for aluminium $K\alpha$, penta-erythritol is far better than potassium hydrogen phthalate, there is little to choose between any of the crystals for the reflection of magnesium $K\alpha$. For both sodium and fluorine, lead stearate is marginally better than potassium hydrogen phthalate. The wavelength range covered by lead stearate is greater than that of the smaller d spacing potassium hydrogen phthalate but against this, potassium hydrogen phthalate is easier to produce and is probably the more stable. Both crystals give rise to crystal fluorescence but for light element work the high energy potassium K radiation is easier to remove by pulse height selection than the M radiation of lead.

TABLE 2.3

Comparison of intensities obtained with lead stearate (L.S.), potassium hydrogen phthalate (K.A.P.), penta-erythritol (P.E.), gypsum and ammonium dihydrogen phosphate (A.D.P.).

Radiation ($K\alpha$)	Sample	Crystal	Peak c/s	Background c/s
Al	Al	P.E.	180,000	
Al	Al	K.A.P.	86,500	
Al	Al	Gypsum	50,000	
Mg	Mg	A.D.P.	27,500	
Mg	Mg	K.A.P.	27,800	
Mg	Mg	L.S.	25,200	
Mg	Mg	Gypsum	20,000	
Na	NaCl	K.A.P.	2,750	9
Na	NaCl	L.S.	3,900	41
Na	NaCl	Gypsum	2,700	35
F	NaF	K.A.P.	344	12
F	NaF	L.S.	782	13

REFERENCES

1. HOLLIDAY, J. E., 1960, Rev. Sci. Instr. **31**, 891.
2. TANEMURA, T., 1961, Rev. Sci. Instr., **32**, 364.
3. MECKE, P., 1963, Z. Analyt, Chem., **193**, 241.
 * 4. SPIELBERG, N., PARRISH, W. and LOWITZSCH, K., 1959, Spectrochim. Acta, **13**, 564.
5. JOHANN, H. H., 1931, Z. Physik, **69**, 185.
6. JOHANSSON, T., 1933, Z. Physik, **82**, 507.
7. ELION, H. A. and OGILVIE, R. E., 1962, Rev. Sci. Instr., **33**, 753.
8. CAUCHOIS, Y., 1932, J. Phys. Radium, **3**, 320.
9. BIRKS, L. S. and BROOKS, E. J., 1955, Analyt. Chem., **27**, 1147.
 * 10. LIEBHAFSKY, PFEIFFER, WINSLOW and ZEMANY, X-*Ray Absorption and Emission in Analytical Chemistry*, Wiley, New York, 1960, Ch. 4.
 * 11. BIRKS, X-*Ray Spectrochemical Analysis*, Interscience, New York, 1959, Ch. 3.
 * 12. BIRKS, *Electron Probe Microanalysis*, Interscience, New York, 1963, Ch. 6.
 * 13. WYTZES, S. A., 1961, Philips Research Reports, **16**, 201.
14. SOLLER, W., 1924, Phys. Rev., **24**, 158.
 * 15. BUWALDA, J., 1964, Philips Serving Science and Industry, **10**, 22.
16. WYTZES, S. A., Philips Technical Review, **27**,11.
 * 17. KLUG, ALEXANDER, X-*Ray Diffraction Procedures*, Wiley, New York, 1954, Ch. 3.
18. ROSE, H. J., ADLER, J. and FLANAGAN, F. J., 1963, Appl. Spectroscopy, **17**, 81.
19. PARRISH, W., *Advances in* X-*Ray Analysis*, Plenum, New York, 1964, 118.
 * 20. SPIELBERG, N. and LADELL, J., 1960, *J. Appl. Phys.*, **31**, 1659.
21. SPIELBERG, N., 1965, Rev. Sci, Instr., **36**, 1377.
22. EBERT, F. and WAGNER, A., 1957, Z. Metalk., **48**, 616.
23. MACDONALD, G. L., *Proceedings of 4th M.E.L. Conference on* X-*Ray Analysis*, (Sheffield, 1964), Philips, Eindhoven, 11.
24. BIRKS, L. S. and SIOMKAJLO, J. M., 1962, Spectrochim. Acta, **18**, 363.
25. FISCHER, D. W. and BAUN, W. L., 1964, U.S. Technical Documentary Report, No. RTD-TDR-63-4232.
26. DINKLAGE, J. and FRERICKS, R., 1963, J. Appl. Phys., **34**, 1633.
27. LANGMUIR, F., 1939, Proc. Roy, Soc., **170A**, 1.
28. BAUN, W. L. and FISCHER, D. W., 1963, U.S. Technical Documentary Report, No. ASD-TDR-63-310.
29. FRANKS, *Proceedings of 3rd International Symposium on* X-*Ray Optics and* X-*Ray Microanalysis*, Academic Press, New York, 1963, p. 199.
30. SAYCE, L. A. and FRANKS, A., 1964, Proc. Roy. Soc., **282A**, 353.

CHAPTER 3

DETECTION

3.1 General[1]

The basic problem of X-ray detection is that of converting the X-rays into a form of energy which can be measured and integrated over a finite period of time. There are numerous ways of doing this and each method depends upon the ability of X-rays to ionise matter. The fundamental difference between the various classes of detector is the subsequent fate of the electrons which are produced by the ionisation process.

The simplest type of detector is a photographic plate in which the photochemical action of X-rays reduces silver halides to free silver via an ionisation process. The total intensity of X-rays will determine the number of silver atoms produced which in turn can be estimated from the blackening of the film.

In the gas filled detectors such as the Geiger counter and the various types of gas proportional counter, the electrons produced by the action of the X-rays are completely removed from the inert gas atom, this being possible in gaseous media because of the relatively long mean free path of the electron. Recombination is prevented by means of a large potential difference across the volume of ionisation and this also enables the electrons to be collected at the anode. Here the current produced forms voltage pulses which can then be amplified and integrated.

There are two types of detector which depend upon solid state electron transfer, these being the scintillation counter and the semiconductor conduction counter. In the scintillation counter the incident X-ray energy promotes valence band electrons in a suitable phosphor to a higher level. When the electrons revert to their original position energy is re-emitted at a longer wavelength which can then be converted into voltage pulses by means of a detector which is sensitive to this range, usually a photomultiplier. In the semiconductor the electrons produced by X-ray bombardment are promoted into conduction bands, the subsequent variation in conduction properties can again be related to the number of incident X-ray quanta.

Of these four main detector categories the second and third are presently the most important and it is the intention to restrict discussion to cover just gas and scintillation counters. Of the other two types, film techniques are invariably avoided because direct integration measurements are not possible. The degree of film blackening has to be measured by means of a photometer, the operation of which can be time consuming and subject to considerable error. Although the potential of semiconductor conduction counters has only recently been recognised,[2-4] the application of these devices has increased rapidly over the last two years in the field of particle spectroscopy. Very recently the use of germanium and silicon semiconductors at low temperatures, has considerably enhanced the scope of gamma-ray spectrometry. A fundamental advantage of the semiconductor counter is that the energy conversion process is very efficient, something like 3 eV only are required to produce a hole/electron pair, compared to around 30 eV for the average energy to produce an ion pair from an inert gas and about 50 eV to effectively produce one light photon from a thallium activated sodium iodide phosphor. As a result the energy resolution of the semiconductor counter is potentially far better than that of the gas proportional or scintillation counters. A basic disadvantage is that the mobility of the doping element at normal temperatures is high enough to cause a considerable deterioration in detector characteristics and as a result the temperature at which these devices have to be operated is necessarily very low. Work is presently in hand investigating the application of these devices for the detection of X-rays. Thin layers of silicon seem to present a suitable detector for X-rays softer than 30 keV, for harder radiation germanium offers better possibilities. The lower limit seems to be approx. 6-8 keV, where a halfwidth value of 1.5 keV has been reported.[5]

3.2 Gas filled detectors[6]

When an X-ray photon interacts with an inert gas atom, an outer electron may be removed leaving a positive ion. For example, in the case of argon —

$$Ar \xrightarrow{\ h\nu\ } Ar^+ + e$$

The resulting combination of electron plus positive ion is called an ion pair. The potential required to remove the first electron from an inert gas atom is fairly small — of the order of 20 eV, the actual value being dependent upon the atomic number of the atom. It will be seen from Table 3.1 that there is a gradual decrease in the values of the first ionization potentials with increase of atomic number, due to the greater shielding of the nucleus by the successive

TABLE 3.1

First Ionization Potentials

Element	Density (g.p.l.)	Atomic Number	Theoretical Ionisation Potential (eV)	Average energy to produce ion pairs (eV)
He	0.179	2	24.6	27.8
Ne	0.900	10	21.6	27.4
Ar	1.874	18	15.8	26.4
Kr	3.708	36	14.0	22.8
Xe	5.841	54	12.1	20.8

addition of principal electron orbits. Owing to the finite range of particles the average energy required to produce an ion pair i.e. the effective ionization potential, is larger than the theoretical first ionization potential. These values have been measured[7] and are also listed in Table 3.1. As the range of a particle decreases with increase of atomic number, the relative difference between the theoretical and effective ionization potentials increases with atomic number. The number (n) of primary ion pairs produced by X-ray ionisation will be proportional to the energy (E_0) of incident X-radiation and will be given by:

$$n = \frac{E_0}{e_i} \tag{3.1}$$

where e_i is the effective ionization potential. After an ion pair is produced rapid recombination will follow unless the electrons are removed from the area of influence of the positively charged atoms. This is achieved by maintaining a potential difference across the volume of the ionized gas.

In its simplest form the gas filled detector consists of a hollow cylinder carrying a thin wire along its radial axis. The wire forms the anode of the detector and carries a potential of around 1-2 keV. The cylindrical casing of the detector is earthed and is filled with a suitable gas mixture, at atmospheric pressure. A window is fitted, either in the wall, or the end of the casing, through which the incident X-rays may pass. An X-ray photon entering the detector produces n primary ion pairs in the manner previously described, but subject to a certain probability, this being dependent upon the probability of absorption of the photon by the counter gas. Provided that the potential at the anode is high enough to prevent recombination, the electrons will move towards the anode and the positive ions to the earthed casing. As the mean free path of an electron is relatively short at normal temperature and pressure, for example, in the case of argon of the order of 4 × 10⁻⁵ cm,[8] the primary electrons soon collide with other inert gas atoms. At low potential differences, these collisions are inelastic and new electrons are produced at each collision leaving the primary electrons in their place. This process continues and results in n electrons eventually reaching the anode. At high potential differences the collisions are no longer inelastic because the kinetic energy gained in one electron mean free path is greater than the energy loss due to collision. Thus a primary electron striking an inert gas atom ionizes the atom without being absorbed, so one electron becomes two, two electrons become four and so on. As the electrons approach the high field region adjacent to the anode wire, the energy gained per mean free path increases considerably and the multiplication process grows rapidly producing the familiar avalanche phenomena.[9] The net effect is called gas amplification or gas gain and its magnitude is given by:

$$G = \frac{N \times e_i}{E_0} \tag{3.2}$$

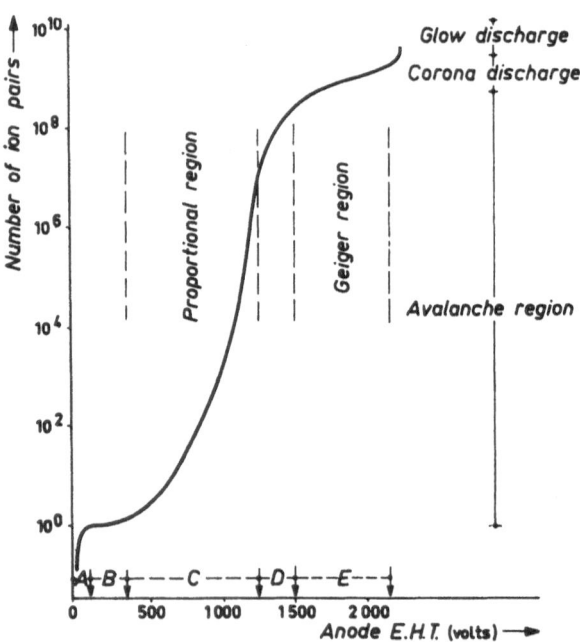

Fig. 3.1 Plot of the number (N) of ion pairs produced at different applied voltages, for a coaxial type gas counter, from a wavelength giving (n) primary ion pairs. As $N = n \times G$, the ordinate represents the effective gas gain (G).

where G is the gas amplification and N the number of ion pairs yielding electrons which eventually reach the anode wire.

Fig. 3.1 shows a plot of N against E.H.T. on the anode for the coaxial cylinder type of gas counter described. In region A the potential difference is insufficient to prevent recombination of the ion pairs and the total gain is less than one. As the potential is increased a stage is reached where just sufficient potential is applied to counteract recombination i.e. $n = N$. This region, labelled B, is called the ion chamber region and is characterised by the fact that the energy of the incident radiation fixes the pulse height and the gain is unity.

As the potential is increased further, gas multiplication starts by a series of individual avalanches and gains of up to 10^6 are achieved. This area C is known as the proportional region and here the localised avalanches start only a few wire diameters from the anode. This region is typified by the fact that the pulse height is proportional to the energy of the incident radiation.

Further increase in the potential causes the individual avalanches to become associated, owing to the production of further electron initators arising from the interaction of UV quanta and the cathode, and a continuous discharge spreads along the full length of the anode. As the ionized electrons are about a thousand times more mobile than the positive ions a complete ion sheath rapidly builds up around the anode and all discharge pulses attain equal size. This space charge effect represents the upper limit of the proportional region which is then followed by a region of limited proportionality D.

Region E is the Geiger region and is characterised by the pulse amplitudes being independent of the energy of the individual X-ray photons. A total gain of 10^9 is possible in the Geiger region.

Even higher potentials bring about corona discharge which continues until the applied potential is removed. Eventually a stage is reached where the current carried by the gas rises rapidly and the voltage drops. This last region is called the glow discharge region.

Fig. 3.2 shows the basic circuitry requirements for the gas counter. The electrons reaching the anode produce a current which is held by a resistor R, in series with the detector anode. This current causes the voltage to drop momentarily producing a pulse which is passed to the cathode follower. In order to produce a pulse of sufficiently high amplitude from a current which may only be of the order of a few microamps, the value of the resistor R is necessarily high, usually of the order of megohms. The impedance of the circuit is therefore very high and the purpose of the cathode follower is to produce a similar pulse under conditions of low impedance. By feeding the pulse to the grid of a triode a similar pulse can be formed at r and low

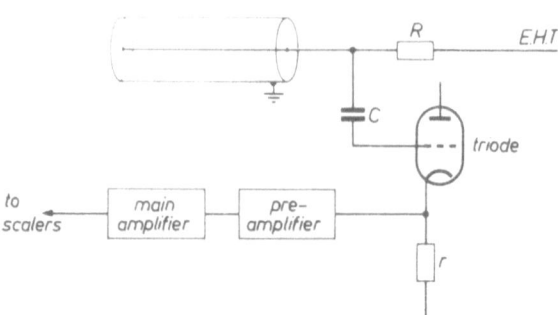

Fig. 3.2 The basic circuitry requirement for the gas proportional counter. In the case of the Geiger counter the pulses from the counter do not require further amplification.

impedance conditions can be achieved by selection of a suitable low value for r — usually of the order of a few hundred ohms. The amplitude of the pulse at R will still determine the pulse amplitude at r although there will be a small constant attenuation by the cathode follower. If the resultant pulse is sufficiently large e.g. as in the case of the Geiger counter, it can be passed straight to the scaling circuits. In the case of the proportional counter, however, some preamplification is necessary.

The fate of the positive ions should now be considered. When the ions reach the cathode, electrons can be transferred from the cathode surface, neutralising the ions, but in turn producing a different photon, which has an energy proportional to the difference in energy between impinging ion and the work function of the cathode surface. As these recombination photons can eject further electrons from the cathode and initiate new avalanches, it is important that their formation be prevented. In the types of gas detector used in X-ray spectrometry it is usual to add a second gas to the ionizable gas which performs the function of a quench, to remove the positive ions. This quench gas is either a compound of a diatomic element, or a hydrocarbon which will produce free radicals fairly easily. These free radicals can donate electrons to the inert gas ions, producing inert gas atoms plus some other quench gas species. Halogens, methane and alkyl halides are frequently used as quench gases. There are three types of gas counter presently in use in X-ray spectrometry, these being the Geiger Müller counter, the sealed proportional counter and the gas flow proportional counter.

3.2.1 DEAD TIME

The relatively slow dissipation of the positive ion sheath has a very significant effect upon the functioning of the counter in that as long as the ions are in the immediate vicinity of the anode, the field is reduced thus preventing further avalanches. This gives rise to the so called dead-time of the counter. The pulse shaping circuitry also contributes significantly to the overall dead time value t seconds, and in modern fast detectors and scaling circuits typical values for t are of the order of 2 micro-seconds. The actual value of the dead time can be estimated since as the counter is effectively dead during the time required for pulse decay, any photon arriving in this period will not be counted. Thus if the time required for the decay of a pulse is t, and R_m pulses arrive in one second, the counter will not have counted pulses for the interval $1-R_m t$ seconds. Thus if it can be assumed that the arrival of pulses follows a purely random distribution the true count rate will be reduced by the fraction $R_m/(1-R_m t)$, where R_m is the measured counting rate. This gives rise to the usual expression for the true counting rate R_t first proposed by Ruark and Brammer. [34])

$$R_t = \frac{R_m}{1-R_m t} \qquad (3.3)$$

One of the assumptions in the derivation of this expression is that the arrival of pulses follows a truly random manner. In practice it is found that pulses tend to arrive in bursts and Equation (3.3) is at best a rough approximation. In fact, provided that t is small the expression is generally applicable to the count rate range normally encountered in quantitative spectrometry, that is, of the order of 50,000 c/s. For example, for a value of t equal to 2 microseconds a loss of 10% would be found over this particular range.

It may be as well to point out, however, that with the advent of higher powered X-ray tubes and faster scalers count rates well in excess of 10^5 c s are likely to become common practice, at such time the use of a more accurate expression for dead time than that given in Eq. (3.3) will be required.

Slightly more accuracy can be obtained by adding more terms to the divisor of Eq. (3.3) which in point of fact should be in the form of a binominal expansion i.e. $1 + R_m t + R_m^2. t^2/2! + R_m^3. t^3/3! + \ldots$ For example at a count rate of 10^5 c/s the first term gives a correction of 20%, the second 2% and so on.

Various methods are available for the measurement of dead time of which probably the ratio method is the best. There are several versions of this method and one of the easiest to apply in practice is that which utilises the ratio of intensities of the $K\beta$ to $K\alpha$ radiation from the same element in a sample containing a certain concentration of the appropriate element. A graphical plot is made of [$\log_{10}$ intensity Kα] versus [intensity Kβ divided by intensity Kα]. Different intensity values are obtained by varying the X-ray tube current and the dead time is calculated by comparing the deviation of the calibration curve from a straight line at the extreme of the required count rate range. The accuracy of this method depends to a large extent on the ability of the goniometer to reproduce the angular position very accurately and should any doubt be possible about this parameter an alternative method should be employed; a survey of various methods has recently been given by Short [35]).

3.2.2 THE GEIGER-MULLER COUNTER[9-10])

The Geiger Müller counter usually called simply the Geiger counter was first developed in 1928.[11]) It is the simplest of the gas counters and has the advantages of being cheap, robust and requires the minimum of ancillary equipment. This last fact is due to its high gas amplification which results in output pulses of the order of a few volts, these being large enough to feed straight into the scaling circuits. Unfortunately the Geiger counter has the fundamental disadvantage of a long dead time which makes it useless for measurement of the large count rates obtainable from modern high power X-ray generators.

Although methods have been described by which Geiger counters can be used at count rates in excess of 10^4 c/s with an associated dead time loss below 20%, [12-13]) the advent of the sealed and flow proportional counters with the combined advantages of small dead time, pulse proportionality and greater specificity is rapidly rendering the Geiger counter obsolete as far as X-ray spectrometry is concerned.

3.2.3 THE PROPORTIONAL COUNTER [14-17])

The proportional counter offers two special advantages over the Geiger counter. The first of these is that the dead time is short, of the order of 0.2 microseconds which is significantly smaller than the scaling circuit dead time (usually about 2 microseconds), thus allowing count rates of up to 5×10^4 to be measured without serious dead time loss. As will be seen from equation (3.3) the dead time loss at 10^5 c/s is only 17% at 2 microsecond dead time. The second advantage is that the mean pulse amplitude is directly proportional to the energy of the incident X-ray photon, thus allowing the elimination of unwanted pulses by use of energy discrimination methods in the form of pulse height selection. Against this the gas amplification of the proportional counter is much smaller than that of the Geiger counter and the associated amplification system has to be far more complex.

The pulse formation process is identical with that of the Geiger counter but as the gas amplification is lower by a factor of about a thousand, the current produced will be of the order of only a few microamps. Thus with a cathode follower R value of around 10 megohms the voltage pulses produced will be of the order of only millivolts. In order to avoid further reduction in pulse size by the capacitance of cables connecting the amplifier and scaling circuits, it is necessary to use a preamplifier to give a pulse amplitude gain of about 10. It is essential to keep the distance between the cathode follower and the preamplifier as short as possible and in sealed proportional counters the two are invariably mounted together with the counter tube. The proportional characteristics of this counter have made it extremely attractive for use in X-ray fluorescence spectrometry, where wavelength harmonic overlap is a common occurrence. Application of pulse height selection methods allow considerable improvement in peak to background ratio and in many cases complete removal of harmonics is possible.

The method of pulse height selection is discussed in detail in the following chapter, but at this stage it is useful to evaluate the pulse characteristics of the proportional counter. Taking a simple case of harmonic overlap, if two radiations of differing wavelength, say SiKα and fourth order FeKβ are incident on a proportional counter, two different size pulses will occur. If a plot of count rate against pulse amplitude is prepared a distribution similar to that shown in Fig. 3.3 will be seen. The pulse amplitude distribution will be seen to consist of three distinct areas, the first of these lies between zero and E_n and consists of noise arising from general electron vibration in the amplifier circuits. The other two areas are symmetrically spread around pulse amplitudes of E_1 and E_2 and correspond to the pulse amplitude distributions due to SiKα and FeKβ respectively.

The general expression for the amplitude of a pulse V_i corresponding to a wavelength λ_i of energy E_i is given by —

$$V_i = \frac{E_i}{e_i} \times A_g \times A_e \times \frac{1}{L} \tag{3.4}$$

where A_g and A_e are the gas and electronic amplification factors respectively. L is the sum of the attenuation stages which may be introduced in order to bring the pulse sizes within the range of the scaling circuits. For a given proportional counter working at fixed E.H.T. A_g is constant, so provided that A_e and L are not varied, V_i will be proportional to E_i, or inversely

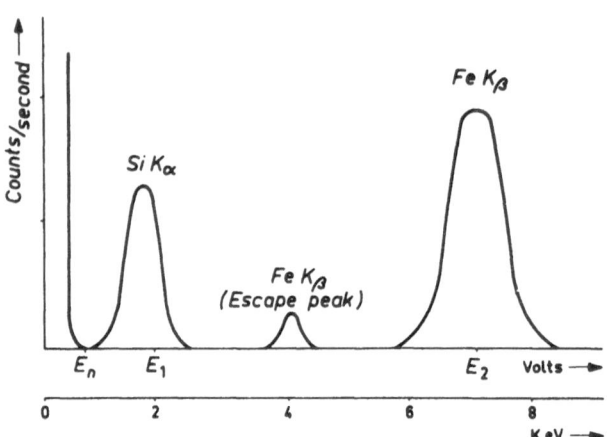

Fig. 3.3 Pulse amplitude distribution from a proportional counter showing energy distributions corresponding to SiKα (1.74 keV) and Fe K_β (7.07 keV).

proportional to λ_t. Thus the values of V_1 and V_2 will be in the inverse ratio

of the wavelengths of SiKα and FeKβ i.e. $\dfrac{1.757}{7.125}$ which equals approximately 0.25.

One immediately noticable feature of the characteristic pulse amplitudes is that they are energy distributions rather than discrete values, this fact being due to the statistical spread in the number of ionizing collisions made by an electron travelling towards the anode.[18] Assuming the distribution of a characteristic pulse amplitude to be gaussian the standard deviation of the curve (σ) will equal $\sqrt{n}$ and the ratio of the peak width at half height (W) to the voltage of the pulse amplitude maximum (V) will equal the resolution (R) of the counter and provided that the anode wire is surrounded by a symmetrical field R will approximate to W/V.

The theoretical resolution of the counter can be expressed in terms of the energy of the incident radiation thus:

$$R = 2.36\sigma$$

since $\sigma\% = \dfrac{100}{\sqrt{n}}$ and from Equation (3.1) for an argon counter $n = E/0.0264$

$$R = \frac{38.3}{\sqrt{E}} \tag{3.5}$$

The assumption that the distribution is Gaussian only holds when n is relatively large (say greater than 50). As a result it is difficult to accurately predict the theoretical resolution of the counter for long wavelength radiation and experimental data has to be used exclusively.

In the cases of SiKα and FeKβ:

	λ	E(keV)	n	R(%)
SiKα	7.125	1.74	66	29.0
FeKβ	1.757	7.07	268	14.3

As will be seen later the usefulness of pulse height selection methods depends upon the relative separation of the pulse amplitude distributions of the interfering wavelengths and the effectiveness of this will obviously depend upon the counter resolution. In order to obtain maximum resolution it is vital that the field surrounding the anode wire be symmetrical and it is for this reason that it is common practice to employ side window counters in order to avoid end effects of the anode wire. End effects can be further re-duced by use of an applied compensating potential to field tubes which are concentrically mounted at either end of the anode wire.[19-20]

Inevitably the side window counter has a relatively short active path length and if used as the solitary detection device, it becomes necessary to use a gas filling with a sufficiently high mass absorption coefficient to absorb as much as possible of the incident radiation.[21] The overall quantum count-ing efficiency will be dependent on the relative absorption efficiencies of the counter filling and the counter window. Xenon and krypton are gases which are frequently used in combination with a window material having the highest possible transmission for the required X-ray wavelength range. However in X-ray spectrometry it is usual to employ a proportional counter with a second window diametrically opposite to the first, such that the radiation which passes through the gas filling can also leave the counter with the minimum of interaction with the walls. When this method is employed a second counter which is sensitive to the harder radiation is placed behind the exit window of the first counter and provision is made to accept either the integrated count rate from both counters, or individual count rates from either. A scintillation counter is invariably used as the second counter. This tandem system offers a considerable advantage of allowing a low mass absorption coefficient gas to be used in the proportional counter, which then becomes specifically sensitive to soft radiation. Thus by use of a low atomic number gas such as argon, the counter filling shows a considerable degree of gas discrimination against short wavelength radiation.

One feature of all proportional counters is the appearance of an additional pulse of lower amplitude than that of the natural pulse, when the incident radiation has a wavelength just shorter than that the of inert gas absorption edge. This additional pulse is called an escape peak and arises wherever the incident radiation can excite characteristic radiation from the inert gas. (see Fig. 3.3) Since all atoms have a very low absorption for their own characteristic radiation, when such radiation is produced it escapes from the counter and an extra pulse is produced having an amplitude corresponding to the difference in energy between the incident radiation and the energy of the characteristic radiation of the inert gas. The proportion of the incident radiation which can excite characteristic inert gas radiation will depend upon the fluorescent yield of the counter gas and its absorption coefficient for the incident radiation. Escape peak to natural peak ratios for argon usually lie between 0.05 and 0.1. Inasmuch as all pulses above the noise level are counted, both the escape and natural pulses will contribute to the total count rate, however, when pulse height selection is employed the presence of escape peaks can present special difficulties and these are discussed in detail in Chapter 4.

As will be seen from Fig. 3.4 the useful working range of the sealed proportional counter usually lies between 0.5 and 4.0 Å, although the short wavelength limit for one specific counter is fixed mainly by its gas filling. The upper wavelength limit arises from the absorption by the counter window, which in sealed proportional counters is usually constructed of 12-15

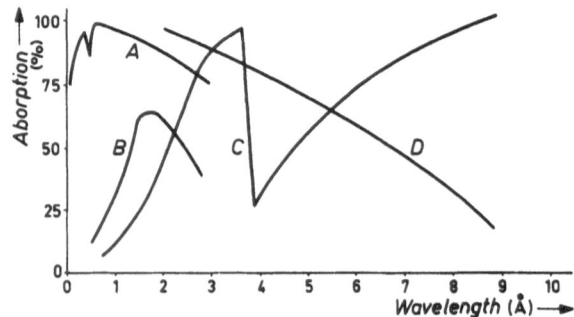

Fig. 3.4 Spectral response curve for the scintillation counter (curve A), the Xe/CH$_4$ proportional counter (curve B) and the Ar/CH$_4$ gas flow proportional counter (curve C). The Ar/Br Geiger counter shows a similar response curve to the proportional counter. Curve D shows the absorption curve for 6 micron mylar film which invariably forms the flow counter window.

micron thickness mica. Use of a window material having a greater long wavelength transmission coefficient invariably results in the window leaking thus rendering the counter useless. This has resulted in the development of the gas flow proportional counter.

3.2.4 THE GAS FLOW PROPORTIONAL COUNTER [22, 23, 36, 37)]

The principle of the gas flow proportional counter is identical with that of the sealed proportional counter and the two differ only in that the gas flow proportional counter is fitted with a very thin entrance window. The most common gas mixture employed in the flow proportional counter is $Ar/10\%$ CH_4 (P.10), but other gas mixtures have been employed particularly in the soft X-ray region where the enhanced short wavelength discrimination of low average atomic number gas mixtures is advantageous. [24)] Among the mixtures have been $Ar/75\%CH_4$(P.75), $He/4\%$ iso-butane and pure methane. The problems due to the possible leakage of gas through the window are overcome by allowing a continuous supply of gas to flow through the detector. Inasmuch as the spectrometer will always be operated under conditions of total vacuum or in a low atomic number inert gas flush, any gas which does leak through the window is immediately removed from the path of the X-radiation. Fig. 3.5 shows the relative absorptions for a variety of window materials. It will be seen that by the use of six micron thickness mylar (polyethylene terephthalate), coated with a few hundred ångström of aluminium to maintain a symmetrical field around the anode wire, the useful range of the counter should extend down to about 10 Å. Confirmation of this is seen in Fig. 3.4). A limit of 10 Å allows the measurement of K radiations

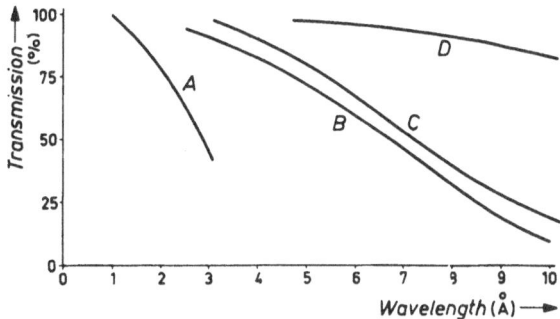

Fig. 3.5 Transmission efficiencies for different counter window materials.

A 3 Mg/cm² Mica   C 25 micron Be
B 6 micron mylar   D 1 micron polypropylene

down to sodium but the range can be increased further by use of ultra thin windows on the counter.[23-25] For example, by use of less than one micron thickness polypropylene flashed with, for example, carbon, measurements have been made down to about 50 Å[26] which includes K radiations down to carbon.

It is of course of vital importance that the window of the flow counter absorbs as little as possible of the incident radiation and improvements in counter techniques over the past couple of years have certainly removed the former doubts about the ability to use ultra-thin windows on a routine basis. Window lives in excess of six months are currently possible with stretched polypropylene provided that care is taken to remove all rough burrs from the contact surfaces of the window. The usual window materials used in the flow counter are as follows:

1 to 8Å   4 or 6 micron mylar or 4 micron polycarbonate
6 to 20Å   1 micron stretched polypropylene
longer than 20Å   collodion supported on a high transmission grid.

For specific problems where it is possible to select analysis conditions for just one element (for example in multichannel spectrometers) it is possible to utilise the poor transmission of a certain type of window material to give some discrimination against unwanted radiations. An example of this is in the determination of fluorine using a teflon $(CF_2)_n$ window in the counter.

TABLE 3.2

Use of counter windows having selective absorption

analysis line	c/s with 1 μm polypropylene	c/s with 6 μm teflon	ratio: polypropylene teflon
MgKα	35,600	1,200	30
NaKα	1,420	9	158
FKα	94	32	2.9

Table 3.2 lists data obtained using windows of 6 micron teflon and 1 micron polypropylene on samples containing magnesium, sodium and fluorine. Due to the favourable position of the fluorine absorption edge for FKα radiation, there is a very large window discrimination against wavelengths

just shorter than the fluorine absorption edge. Thus for high fluorine sensitivity a thin (say 1 micron) teflon window could be used, or alternatively,
where wavelength discrimination is required, a relatively thick teflon window would be the best. This latter point is of great importance in non-
dispersive work where good response to background ratios are difficult
to obtain, and for reasons of inefficiency of pulse height selection it is vital
that radiations immediately next to the energy of the required radiation be
removed by alternative means.

3.3 The scintillation counter

The conversion of X-ray energy into voltage by means of the various types
of gas detector is always a single stage process. In the case of the scintillation
counter however, a double stage process is involved consisting of first
conversion of X-ray energy into a much lower energy state, usually around
2.5-3.0 eV and the second of converting this long wavelength energy into
volts. The first of these two stages is achieved by means of a phosphor and
the second by means of a photomultiplier.

3.3.1 THE PHOSPHOR

A phosphor is a substance which absorbs radiation of a certain wavelength
and then re-emits at a longer wavelength. The simplest phosphors are aro-
matic organic compounds containing a high proportion of conjugated π
bonding — anthracene, naphthacene and pentacene are typical examples.
These molecules are excited by the incident radiation and provided that this
excitation does not bring about irreversible configurational changes, the
molecule will re-emit radiation as the electrons return to the ground state.
Most phosphors used in scintillation counter are inorganic compounds
which are doped with impurity elements. The inorganic compound must have
a fairly low ionization potential and must be preparable in the form of high
purity, large single ionic crystals, of simple crystallographic structure. Certain
alkali halides have all of these properties and are invariably used as the
phosphor base material. The doping element must have a covalent radius
of the same order as the alkali halide atoms and must have a stable mono and
a higher valence state, thallium and europium have both been employed
with considerable success. At the present time nearly all of the phosphors in

use in X-ray spectrometry consist of thallium activated sodium iodide.[27-28] The mechanism of the energy conversion process[10] starts with the excitation of a halide valence band electron into a conduction band leaving a positive hole. The positive hole migrates to the nearest impurity atom centre where the atom is ionized, thus being oxidised to its high valence state. Hence immediately after the primary ionization the phosphor contains electrons in its conduction band and has one positive centre for each electron. Each electron drops into a centre forming an excited state which rapidly decays to the normal ground state with emission of radiation. The decay time is normally very short and is usually of the order of 0.1 of a microsecond.

In the case of thallium activated sodium iodide, the potential (e_i) required to produce one light photon is about 3 eV and the emitted radiation is in the blue region having a wavelength of around 4100 Å. In order to convert this energy further into a current which can be used to produce a voltage pulse, a photomultiplier has to be used which is sensitive to the blue region. The total efficiency of the scintillation counter will depend, not only upon the efficiency of the energy conversion of the light photons by the photocathode but, in addition, on the fraction of incident X-radiation absorbed and the number of light photons which actually reach the photocathode of the photomultiplier. Hence the dimensions and method of mounting of the phosphor are most important. The phosphor must be thick enough to absorb the majority of the incident X-radiation conducive with the property of a high optical transmission for the light photons which are produced and which have to pass through the crystal to the photocathode of the photomultiplier. As sodium iodide is hygroscopic the crystal mount must be both airtight and optically opaque at its outer side, but in addition, it must have good transmission characteristics for X-rays. A crystal thickness of about 2 mm with an outer window material consisting of 0.2 mm of beryllium coated with a micron or so of aluminium, is a satisfactory combination for wavelengths below 3 Å. Finally, it is vital that a good optical coupling be maintained between the emission face of the phosphor and the photomultiplier, and to ensure that this is so, it is usual to have a thin smear of silicone oil between the two optically similar faces.

3.3.2 THE PHOTOMULTIPLIER

The working of the photomultiplier depends upon the principle of secondary emission. If an electron at the surface of a solid material is given energy in excess of the surface energy of the solid, the electron can escape. If this electron can gain sufficient energy in one mean free path, on collision

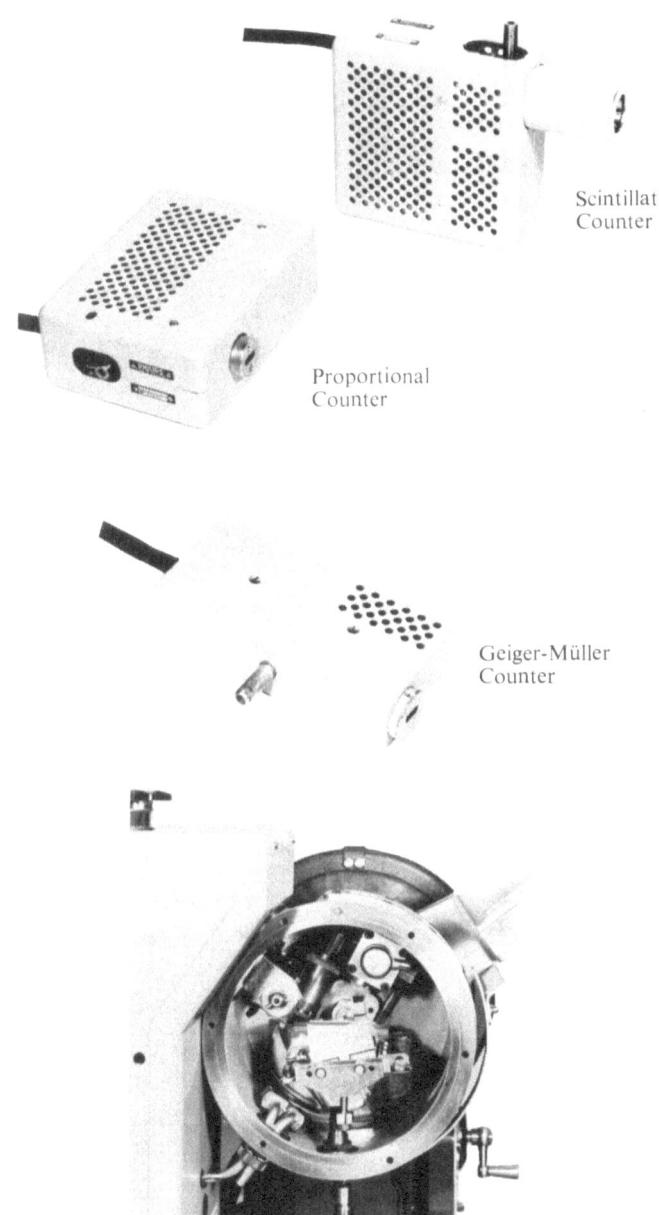

Scintillation
Counter

Proportional
Counter

Geiger-Müller
Counter

Fig. 3.7 Examples of detectors for X-ray spectrometry from the Philips range of equipment. The gas flow counter is shown mounted in its normal position inside the crystal chamber of the spectrometer. The cathode follower can be seen at the back of the chamber to the left of the counter body. The preamplifier is mounted in the bottom of the sample chamber on the left of the picture.

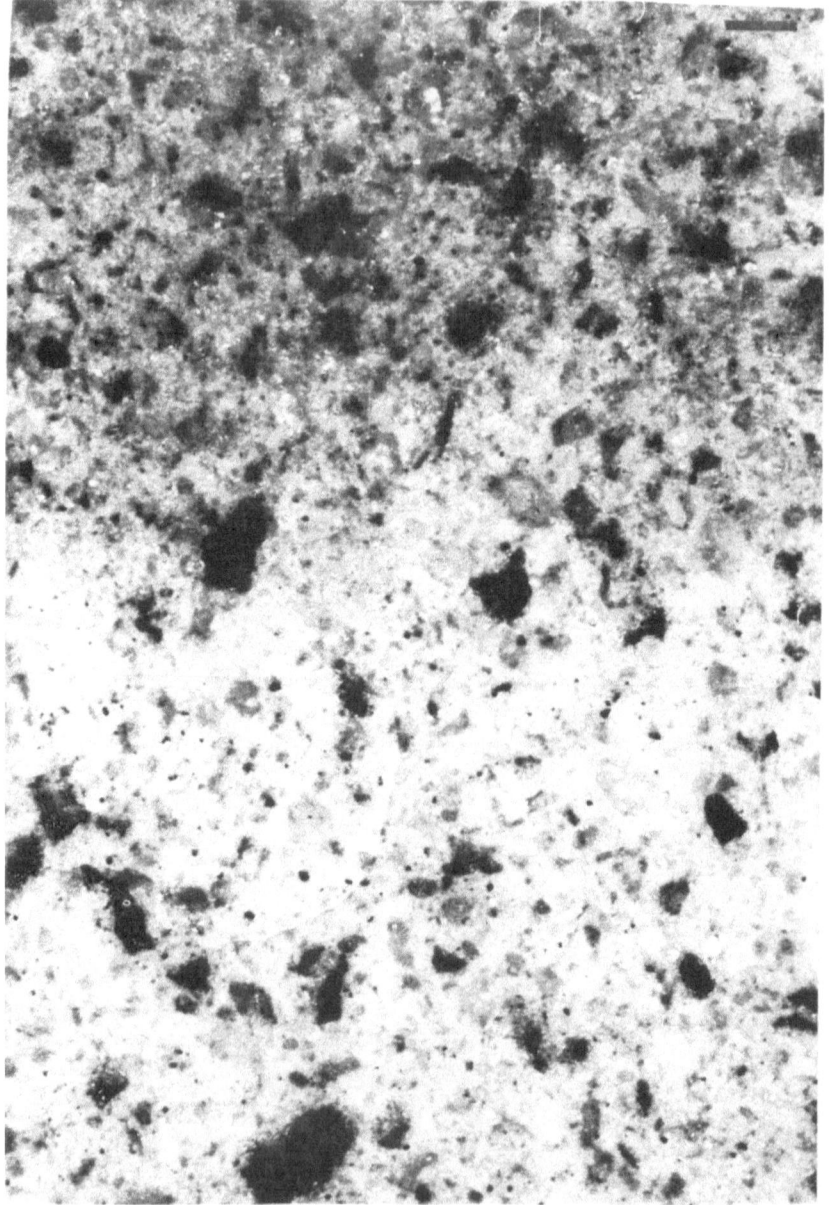

50 µm

Fig. 6.7 Photomicrograph of a synthetic syenite sample after grinding for twelve minutes in a disc mill. The very wide range in particle sizes can be clearly seen including many well in excess of 50 µm. The majority of these larger particles are of mica.

with a photosurface more than one electron may result. To this extent the principle of the photomultiplier is similar to that of the gas counter because in each case an overall gain is achieved by the acceleration of primary electrons. Fig. 3.6 shows an idealised scheme of the photomultiplier tube. The photocathode material is Sb/Cs and when the light photons from the phosphor are incident on its active surface bursts of electrons are produced. These electrons are focussed electrostatically onto the first of a series of 10 dynodes each of which has a successively higher, applied positive potential. When the electron strikes the first dynode a number of secondary electrons are produced. These electrons are then further accelerated to the second dynode where an even larger number of electrons result. The process continues until the final electron output is collected at the anode, where a small current is produced and voltage pulses arise in exactly the same way as in the gas

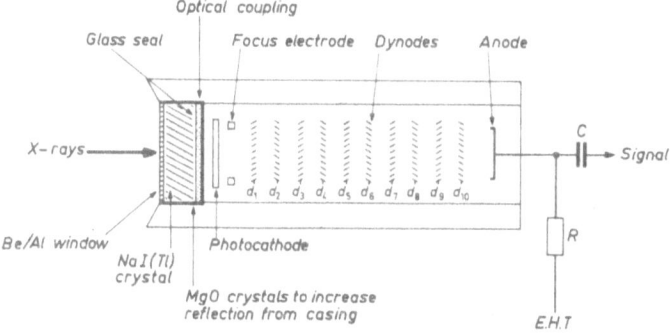

Fig. 3.6 The scintillation counter. The X-rays are converted into light energy by the NaI(Tl) Phosphor. This light falls onto the Sb/Cs photocathode producing a burst of electrons which are focussed onto the first of a series of 10 dynodes (d_1—d_{10}). Each dynode has a succesively higher positive potential and the electrons produced at each dynode are accelerated to the next, until they eventually reach the anode. Here a small current is produced causing a momentary drop in voltage and each pulse is passed via the condenser C to the cathode follower and amplifying circuits on to the scaling circuits.

counters. If the number of secondary electrons produced at the first dynode is δ the total gain of the photomultiplier will equal δ^m where m is the number of dynodes. For energies falling within the blue region i.e. about 3 eV, δ has a value of about 4, thus for a 10 stage photomultiplier tube the total gain is 4^{10} which equals approximately 10^6.

Several applications have been reported in which special photomultipliers have been used directly in the measurement of soft X-radiation.[29-33] These have been mainly Cu/Be photomultipliers which have to be operated under conditions of high vacuum. Since their optimum sensitivity range is usually in excess of 50 Å, to date, they have found little application in conventional X-ray spectrometry. However, their potential usefulness is being greatly enhanced by the continuing development of soft X-ray spectroscopy.

3.3.3 CHARACTERISTICS OF THE SCINTILLATION COUNTER

The gain of the scintillation counter is of the same order as the gas gain of the proportional counter, but as the ionization potential required to effectively produce one blue photon is almost twice that required to produce one ion pair from an inert gas atom, the amplitude of the pulses produced by the scintillation counter from a certain wavelength, will only be one half that of the equivalent pulse amplitude from the proportional counter. The energy transference process in the scintillation counter is rather inefficient and only about 7 % of the primary photons effectively contribute to the current produced at the anode.[14] Thus the resolution of the scintillation counter is far worse than that of the proportional counter. For example, taking the case of FeKα radiation ($\lambda = 1.937$ Å), since the NaI(Tl) phosphor emits photons at about 4100 Å the number of photons produced will equal $4100/1.937 = 2117$. Assuming a crystal efficiency of 20 % and a photomultiplier efficiency of 5 %, the effective number of photons $n = 21.2$. As in the case of the gas proportional counter, the resolution $R = 2.36\ \sigma$ and $\sigma\% = 100/\sqrt{n}$, thus

$$R = \frac{129}{\sqrt{E}} \tag{3.6}$$

and hence R for FeKα equals approximately 51 % for the scintillation counter compared to about 15 % for the gas flow proportional counter. The significance of this fact will become apparent in the following chapter on pulse height selection.

From the foregoing it is apparent that the poor resolution of the scintillation counter in the X-ray region is due mainly to the relatively large potential required to produce a light photon from the phosphor and to the poor efficiency of the photocathode. Recent developments in phosphors[1] have indicated that CaI_2/Eu should yield energy resolutions around 30 % in the 6 keV region.

An unique advantage of the scintillation counter is its almost constant spectral response over a very large range, Fig. 3.4. Its useful working range is limited at the long wavelength end of the spectrum by the relative large amplitude of the noise pulses from the photomultiplier and by the absorption of the soft radiation by the casing of the phosphor. In effect this means that the scintillation counter would not normally be used for K radiations from elements of lower atomic number than titanium (22) or for L radiations from elements of lower atomic number than iodine (53). Escape peaks can occur also in the scintillation counter following the production of iodine K or L radiations from the sodium iodide in the phosphor. However, since radiations arising from the iodine L edges are around 5 keV, which is outside the normal working range of the scintillation counter, only radiation with energy in excess of the iodine K edge (33 keV) will give rise to escape peaks.

3.4 Comparison of detectors

Fig. 3.7 shows typical commercial examples of the four types of detector discussed. In the proportional and scintillation counters the cathode follower/ pre-amplifier is mounted along with the counter tube, as is the case with the cathode follower for the Geiger tube. As the gas flow counter is normally mounted inside the spectrometer, the counter and its cathode follower are arranged in such a way as to overcome the space limitations inside the body of the spectrometer. Table 3.3 lists the more important of the characteristics of the Geiger, proportional, gas flow and scintillation counters.

TABLE 3.3

Characteristics of Counters

	Geiger	Proportional	Gas Flow	Scintillation
Window	Mica	Mica	Mylar/Al	Be/Al
Thickness	3 mg/cm^2	2.5 mg/cm^2	6 μm	0.2 mm
Position	Radial	Axial	Axial	Radial
Filling	Ar/Br	Xe/CH$_4$	Ar/CH$_4$	—
Pre-amplifier	unnecessary	x10	x10	x10
Auto-amplification	10^9	10^6	10^6	10^6
Useful range (Å)	0.5 — 4	0.5 — 4	0.7 — 10*)	0.1 — 3
Dead Time (Micro seconds)	200	0.5	0.5	0.2
Max. useful count rate	2 / 10^3	5 / 10^4	5 / 10^4	10^5
Cosmic background (c/s)	0.8	0.4	0.2	10
Resolution % (FeKα)	—	14	15	51
Quantum counting efficiency	λ dependent	λ dependent	λ dependent	reasonably independent of λ

*) Can be extended to 50 Å by use of ultra-thin windows.

Briefly, the Geiger counter has the advantage of simplicity and requires the minimum of amplification equipment. Against this, it has a large dead time and does not show proportional characteristics. Conversely, the proportional and gas flow counters have the advantage of proportionality and

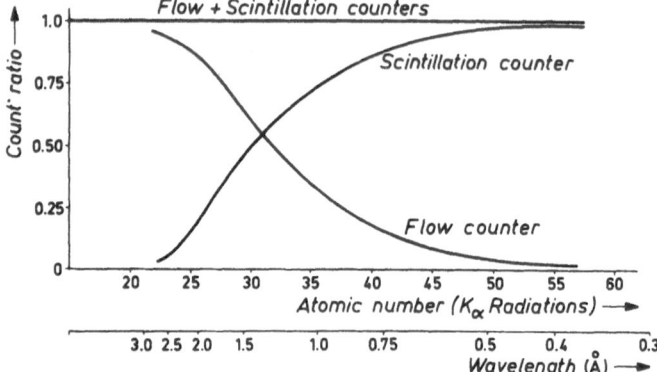

Fig. 3.8 Comparison of the count rates obtained individually from the flow proportional counter and the scintillation counter, compared with the count rates obtained from the same counters in series.

short dead time but require far more sophisticated ancillary equipment. In addition, both have good energy resolution but the gas flow counter can require much attention if its resolution is to be maintained. The scintillation counter has the added advantage of constant quantum counting efficiency over a large wavelength range, but has poor resolution and shows relatively high background rates. All of the gas counters have the advantage that a certain degree of gas discrimination against short wavelengths can be achieved by careful selection of gas filling and effective absorption path length. Most of the advantages of both scintillation and gas flow counters can be gained by use of a combination detector consisting of a scintillation counter mounted behind a gas flow counter. Fig. 3.8 shows a plot of relative counting efficiencies against wavelength for a typical tandem detector and it will be seen that the integrated count from the combined detectors is well in excess of count rates from individual detectors, for something like 50 % of the atomic number range. However, the selection of a detector is not normally made on counting efficiency alone and more often than not, optimum peak to background ratio has also to be considered. It is for this reason that the spectroscopist should be furnished with the means of selecting all three possibilities of detector type. (See also Appendix 2a).

REFERENCES

1 KILEY, W. R. and DUNNE, J. A., 1963, A.S.T.M. Spec. Tech. Publ., No. 349, 24.
* 2 Report on 9th Scintillation and Semiconductor Counter Symposium, 1964, Nucleonics, 49.
3 SHIRLEY, D. A., 1965, Nucleonics, 23, 62.
4 MILLER, G. L. and GIBSON, W. M. and DONOVAN, P. F., 1962, Ann. Rev. Nuclear Sci., 12, 189.
5 BOWMAN, H. R., HYDE, E. K., THOMPSON, S. G. and JARED, R. C., 1966, Science, 151, 562
* 6 CURRAN and CRAGGS, Counting Tubes, Butterworths, London, 1949.
7 COMPTON and ALLISON, X-Rays in Theory and Experiment, Van Nostrand, New York, 1936.
8 LONG, A., 1959, J. Brit., I.R.E., 19, 273.
9 FRIEDMAN, H., 1949, Proc. I.R.E., 37, 791.
* 10 SHARPE, Nuclear Radiation Detectors, Methuen, London, 1964.
11 GEIGER, H. and MULLER, W., 1929, Phys, Z., 29, 839.
12 MCKEOWN, P. J. A. and UBBELOHDE, A. R., 1954, J. Sci. Instr., 31, 321.
13 TROST, A., 1950, Z.angew. Phys., 2, 286.
14 LANG, A. R., 1951, Nature, 168, 907.
* 15 LANG, A. R., 1956, J. Sci, Instr., 33, 96.
16 ARNDT, U. W., COATES, W. A. and CRATHORN, A. R., 1954, Proc. Phys. Soc., 67, 357.
* 17 MULVEY, T. and CAMPBELL, A. J., 1958, Brit. J. Appl. Phys., 9, 406.
* 18 CURRAN, S. C., ANGUS, J. and COCKROFT, A. L., 1949, Phil. Mag., 40, 929.
19 BEHN RIGGS, F., 1963, Rev. Sci. Instr., 34, 392.
20 COCKROFT, A. L. and CURRAN, S. C., 1951, Rev. Sci. Instr., 22, 37.
* 22 PARRISH, W. and TAYLOR, J., 1955, Rev. Sci. Instr., 26, 367.
* 22 HENDEE, C. H., FINE, S. and BROWN, W., 1956, Norelco Reporter, 3.
23 HENKE, Advances in X-Ray analysis, Plenum, New York, 1962, 6,
* 24 FISCHER, D. W. and BAUN, W. L., 1964, A.F. Materials Laboratory Report, No. RTD-TDR-63-4232.
25 DUNNE, J. A., 1964, Norelco Reporter, 11, 109.
26 HENKE, Advances in X-ray analysis, Plenum, New York, 1963, 7, 360.
27 NELSON, J. T. and ELLICKSON, R. T., 1955, J. Opt. Soc. Amer.
28 BRETANO, J. C. M. and LADANY, I., 1954, Rev. Sci. Instr., 25, 1028.
29 JACOB, L., NOBLE, R. and YEE, H., 1960, J. Sci. Instr., 37, 460.
30 FRANKS, A., 1964, Nature, 201, 913.
31 DAWBER, K. R., 1960, Quarterly Progress Report, M.I.T. Res. Labs. Electr., No. 65, 62.
32 BEDO, D. E. and TOMBOULIAN, D. H., 1961, Rev. Sci. Instr., 32, 184.
33 TIUTIKOW, A. M. and EFREMOV, A. J., 1958, Soviet Phys. Doklady, 3, 154.
34 RUARK, A and BRAMMER, F. E., 1937, Phys. Rev. 52, 322.
35 SHORT, M. A., 1966, Proc. 5th Conference on X-ray Analytical Methods (Swansea, 1966) Philips, Eindhoven. 60,
36 CAMPBELL, A. J., 1967, Norelco Reporter, 14, 103.
* 37 JENKINS, R., 1968, Philips Scientific Reports, 79.136.FS6, Philips, Eindhoven.

CHAPTER 4

PULSE HEIGHT SELECTION

4.1 Principle of pulse height selection[1-6]

Pulse height selection affords a method of isolating a moderately narrow range of wavelengths from a spectrum by virtue of energy separation as opposed to wavelength separation as, for example, in crystal dispersion. By use of a proportional counter each incident wavelength is converted into a voltage pulse or rather a pulse distribution, of characteristic pulse amplitude and the function of the pulse height selector is to allow only a specific and chosen range of pulse amplitudes to pass to the scaling circuits. The idealised principle of such a system is illustrated in Fig. 4.1.

Three wavelengths λ_1, λ_2 and λ_3 are incident on the counter producing three pulse amplitudes V'_1, V'_2 and V'_3. These pulses are amplified by the linear amplifier in order to make them sufficiently large for the subsequent selection and scaling processes but the ratio of $V'_1 : V'_2 : V'_3$ is maintained.

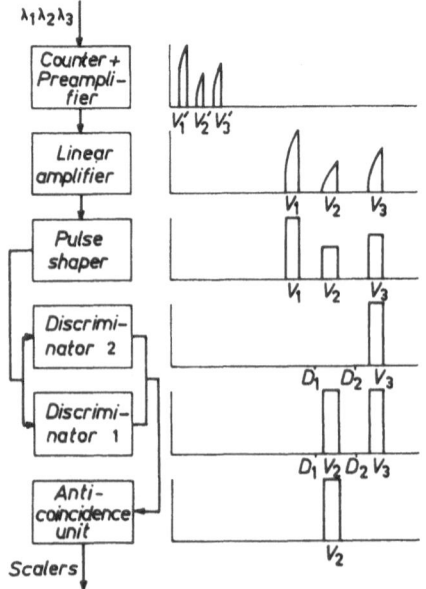

Fig. 4.1 Block diagram of the single channel pulse height selector.

Pulses V_1, V_2 and V_3 from the linear amplifier are passed through the pulse shaper to the discriminators where each discriminator produces a rectangular output pulse, for each input pulse above a pre-selected voltage level. The first discriminator is set at level D_1 and as both V_2 and V_3 are of greater amplitude than this, two corresponding pulses are produced. The second discriminator is set at level D_2 but this time only the pulse corresponding to V_3 is produced. The outputs from both discriminators are fed to an anti-coincidence circuit which is simply a device to prevent the passage of simultaneous pulses from different discriminator levels. The V_3 pulse is common to both levels and so is rejected and in addition, as the V_1 pulse is insufficiently large to pass the first discriminator level, only V_2 is passed to the scaling circuits. Thus the effect of the process is to select pulse heights between the pre-determined discriminator levels D_1 and D_2.

In normal practice, the single channel pulse height selector used in X-ray fluorescence spectrometry is provided with a continuously variable lower level i.e. D_1 and a range of window (channel) width settings; this window width corresponds to the required range of levels i.e. $D_2 - D_1$. It is common practice to fix the electronic amplification such that the lowest energy i.e. the longest wavelength, falls within the range of the pulse height selector and then to provide an attenuator in order to reduce the larger pulses such that they too fall within the working range. Some flexibility is also possible by virtue of the fact that the gas amplification can be varied by changing the E.H.T. on the counter.

4.2 Automatic pulse height selection

Since the mean pulse amplitude is proportional to energy, a different setting of the pulse height selector will be required for each different wavelength to be measured. The re-setting of the pulse height selector can be a tedious process and is a particular disadvantage in automatic X-ray spectrometry where speed is an all important factor. Thus it is apparent that there is a considerable incentive to automate the pulse height selection process and at least one commercially available spectrometer has such a facility. There are several ways of carrying out this automation[5] and these can be illustrated with reference to Fig. 4.2. This shows a typical case occuring in X-ray fluorescence spectrometry where a wavelength giving a pulse amplitude at V_3 and an escape peak at V_1, interferes with an analytical line giving V_2. The rejection of V_1 and V_3 can be achieved in one of two ways — either by bringing V_2 into a preset range of D_1 and D_2, or by setting D_1 and D_2 to embrace a preset value of V_2.

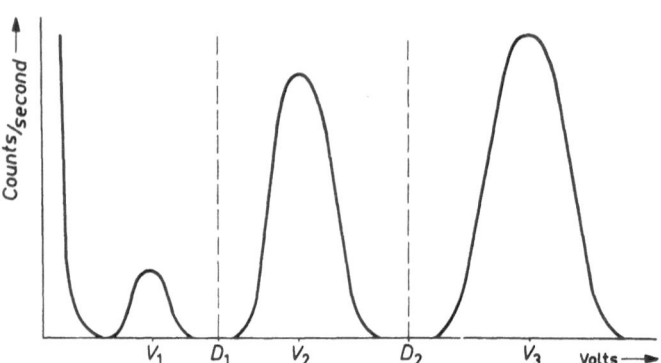

Fig. 4.2 Application of pulse height selection to a simple case of harmonic overlap. The required wavelength λ_2 gives pulses at V_2 which are interfered with by a harmonic V_3 and its escape peak V_1. The wavelength of the harmonic λ_3 will be equal to ($V_3/V_2 \sim \lambda_2$).

4.2.1 VARIATION OF PULSE AMPLITUDE

In this method all pulses are amplified by varying degrees in order to bring them within the range of a fixed and preset channel width. Equation (3.4) shows that this amplification can be brought about either by varying the gas gain A_g (or the photomultiplier gain in the case of the scintillation counter) by changing the detector voltage, or by varying the electronic amplification i.e. $\left(\dfrac{A_e}{L}\right)$. Of these possibilities a change in electronic amplification or attenuation is the more attractive proposition since the gas gain is far more dependent upon the characteristics of the counter. If A_g is kept constant V_i will be proportional to $E_i \times \left(\dfrac{A_e}{L}\right)$. Since E_i is proportional to $\dfrac{1}{\lambda_i}$, and from the Bragg relationship λ_i is proportional to $\dfrac{2d \sin \theta_i}{n}$, it can be seen that:

$$\left(\frac{A_e}{L}\right) \propto V_i \cdot \sin \theta_i \times \frac{2d}{n} \tag{4.1}$$

Thus for a crystal of a fixed $2d$ value, reflecting a certain order, if $\left(\dfrac{A_e}{L}\right)$ is varied sinusoidally with θ all pulses will occur at V_i.[7] This is the system employed in the Philips PW series automatic X-ray spectrometers,[8] where fixed resistor banks are provided in order that the system can be trimmed for first or second order reflections from a wide range of crystals.

One of the disadvantages with this system is that window width remains constant, where as the required window width varies with the energy (E) of the radiation. From Equation (3.5) it can be seen that resolution of a detector is proportional to the $\sqrt{E}$ thus, as E increases, the required window width of the pulse height selector decreases and as will be seen later, under certain circumstances this can introduce extraneous background.

4.2.2 VARIATION OF BASE LINE AND CHANNEL SETTINGS

A second method consisting of varying base line and/or window settings, has also been applied with varying degrees of success. The simplest system entails varying the base line of the discriminator (D_1) linearly with the 2 θ setting of a spectrometer which is linear in λ.[9] A far more sophisticated version of this procedure also varies the window width and this provides a much more ideal system.[10] In principle, this is similar to the pulse amplitude variation method, except that $\dfrac{A_e}{L}$ is now held constant and the condition $D_1 \propto \dfrac{1}{\sin \theta}$ is used to set D_1. As D_1 changes the channel width should theoretically change as $\dfrac{1}{\sqrt{\sin \theta}}$ but it has been shown that in practice a linear change of window width with θ is sufficient.

4.3 Applications of pulse height selection[11]

In principle, it should be possible to rely on the exclusive use of pulse height selection to abstract a wavelength or a range of wavelengths from a complex spectrum. In view of the high quantum loss experienced from inefficiency of the crystal dispersion/collimation process (usually less than 5 %) this appears an attractive proposition but in practice, the usefulness of the non-dispersive method is severely limited by the inherently poor resolution of the detector systems which have to be used.[12 18] Although several applications have been reported in which non-dispersive methods have been employed with,[19 20] and without,[21 23] the added use of filters, it is common practice to use pulse height selection to supplement crystal dispersion. However, the advent of fast multi-channel pulse height analysers now offers the possibility of employing non-dispersive methods based on the sorting of mixtures of pulse amplitude distributions.[24] This procedure of convolution sorting has already been applied with success both to γ-ray spectrometry[25] and to electron probe microanalysis,[26] but to date has not been applied to radiation intensities obtained from conventional sources. For this reason it is the intention to limit the discussion to the use of pulse height selection as an added aid to crystal dispersion.

4.4 Theoretical application of pulse height selection

The success with which pulse height selection can be applied, depends upon the relative energy differences between the pulses required for analysis and those to be rejected; hence three factors are of paramount importance:

a. Differences in energy of the component distributions.
b. Shapes of the pulse amplitude distributions.
c. The background distribution.

Fig. 4.2 can be used to illustrate the first two of these factors following the effect of harmonic overlap. From the Bragg relationship $n\lambda = 2d \sin \theta$, it will be seen that from a specific diffraction angle for a certain crystal of fixed $2d$, the first order of λ_1 will be overlapped by the second order of $\lambda_1/2$ the third order of $\lambda_1/3$ and so on. By calculating the shape of the pulse amplitude distributions from the energy of the radiations and from the known resolution of the detector, it is possible to calculate the degree by which, for example, V_2 and V_3 are resolved. The following worked example should serve to illustrate this calculation.

Example

MnKα is partially overlapped by Au2Lβ_2. Predict the relative efficiencies of the flow counter and the scintillation counter when combined with pulse height selection for the removal of Au2Lβ_2 interference. The measured resolution of the two counters for FeKα radiation is 17 % for the flow counter and 50 % for the scintillation counter.

4.4.1 FLOW COUNTER

Equation (3.5) gives the resolution of a counter in terms of the energy of the incident radiation and this can be expressed in a more general form relating the relative resolutions of the counter for different energies viz:

$$R_1 = R_2 \sqrt{E_2/E_1} \qquad (4.2)$$

Since the given resolution for FeKα (6400 eV) is 17 %, the resolution corresponding to $MnK\alpha$ (5880 eV) and $AuL\beta_2$ (11570 eV) will be 17.8 $\%$ and 12.6 % respectively. As

$$R = \frac{W}{V} \tag{4.3}$$

where W is the peak width at half height of a pulse amplitude distribution peaking at V, an arbitrary value of E_0 for one of the radiations can be assumed for fixed amplification and all calculations based on this. Thus, for example, taking a peak value of 25 volts for $MnK\alpha$, $AuL\beta_2$ will peak at $25 \times \frac{11.57}{5.88} = 49.3$. From equation (4.3) W for $MnK\alpha$ will equal $\frac{17.8}{100} \times 25 = 4.4$ volts and W for $AuL\beta_2 = \frac{12.6}{100} \times 49.3 = 6.1$ volts.

If it is assumed that the pulse amplitude distributions are Gaussian (this is an acceptable approximation but, in fact, the pulse amplitude distributions are slightly distorted to the high energy side), the W value will correspond to 2.36σ. It can be further predicted that 95.4 % of the area under the pulse amplitude distribution will lie between $V \pm 2\sigma$ and 99.7 $\%$ between $V + 3\sigma$. Hence 99.7 % of the pulse amplitude distribution for $MnK\alpha$ will lie between 25 ± 5.5 volts, as

$$\sigma = \frac{4.4}{2.36} = 1.85 \text{ volts.}$$

For $AuL\beta_2$ $\sigma = \frac{6.1}{2.36} = 2.5$ volts, 99.7 % of the total peak area will be confined between 49.3 ± 7.8 volts. As the upper level for $MnK\alpha$ is found to be 30.5 volts and the lower level for $AuL\beta_2$ 41.5 volts, there will be complete resolution between the main peaks. A small residual intensity of the escape peak of $AuL\beta_2$, at $[11.57 - 2.95] \times \frac{25}{5.88} = 36.3$ volt will, however, pass through the Mn-window.

4.4.2 SCINTILLATION COUNTER

The given resolution for $FeK\alpha$ is 50 % and from Equation (4.2) the resolution for $MnK\alpha$ will equal 52.2 % and that for $AuL\beta_2$ 37.2 $\%$. From Equation (4.3) the W value for $MnK\alpha$ will equal $\frac{52.2}{100} \times 25 = 13$ volts and that for $AuL\beta_2$ $\frac{37.2}{100} \times 49.3 = 18.4$ volts. From these data it is obvious that there will be some overlap of the two distributions and the amount of overlap can be estimated from the following table.

MnKα		AuLβ₂	
Distribution	Limit of upper level	Distribution	Limit at lower level
σ 25 ± 5.5 volts	30.5 volts	49.3 ± 7.8 volts	41.5 volts
2σ 25 ± 11.0 volts	36.0 volts	49.3 ± 15.6 volts	33.7 volts
3σ 25 ± 16.5 volts	41.5 volts	49.3 ± 23.4 volts	25.9 volts
4σ 25 ± 22 volts	47.0 volts	49.3 ± 31.2 volts	18.1 volts

The overlap occurs at just below the 2σ level hence the efficiency of a perfectly set pulse height selector in the removal of $Au2L\beta_2$ will be 90 to 95 %.

In practice it is found that about 90 % of $Au2L\beta_2$ can be removed when pulse height selection is applied with the scintillation counter and over 99 % when the flow counter is used. Sometimes it is possible to reduce unwanted radiation by deliberately missetting of the window levels. A setting of 26 volts as upper level and 8.5 volts as lower level would eliminate the Au-radiation almost completely but reduce the Mn-intensity to slightly more than half its maximum value. This asymmetric setting can be used, when the intensity

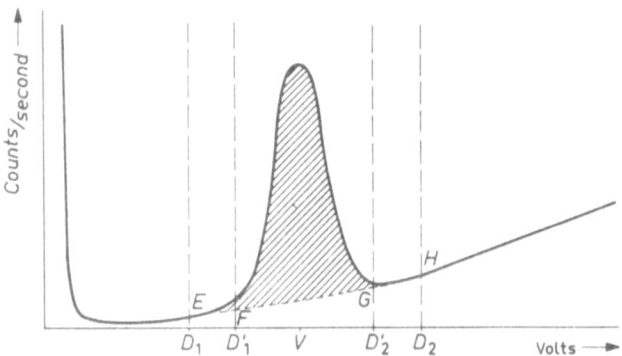

Fig. 4.3 Application of the pulse height selector to a pulse amplitude superimposed upon an increasing background of the same energy range. Careful setting of the discriminator levels can result in considerable improvement in the signal to noise ratio.

of the Au-line is very much higher than that of the Mn-line. However, the stability requirements are very severe if the window limit sits on a steep slope of the pulse distribution.

The effect of the background distribution is illustrated in Fig. 4.3. If the extreme edges of the pulse amplitude distribution do not go down to zero count rate, the setting of the selected window width will be rather critical. In the example shown, a pulse amplitude distribution peaking at V is superimposed on a background which steadily rises towards the high energy end. With the discriminator levels set at D_1 and D_2 the effective peak count rate is proportional to the hatched area under the pulse amplitude distribution. The effective background i.e. "blank" would be similarly proportional to the area bounded by D_1, E, H and D_2. However, by re-setting the discriminator levels to D'_1 and D'_2, although the count corresponding to the hatched area is not significantly effected, the background is now proportional to the area bounded by D'_1, F, G and D'_2 and a considerable decrease in the "blank" count rate will result. This situation is typical of the case where pulse height selection is applied to shorter wavelengths and the background is due mainly to low order scattered continuous radiation. It is clear that the efficiency of this background reduction depends on the wavelength to be measured. The longer this wavelength, the better the harder background intensity can be reduced. The useful effect of P.H.S. in reducing background when using a scintillation counter is thus much greater for wavelength like FeKα than for MoKβ.

4.5 Practical problems arising in pulse height selection

The problems which occur in the application of the pulse height selector arise mainly from shifts or distortions of the measured pulse amplitude distributions and occur mainly when the gas flow proportional counter is used. These effects are illustrated in Fig. 4.4. Fig. 4.4a demonstrates the effect of a pulse amplitude shift, where, with the discriminator levels set at D_1 and D_2 to embrace the pulse amplitude peaking at V, the count rate is proportional to the area under the pulse amplitude distribution. If the pulse amplitude should shift after the setting up procedure, for example, to V', the total count rate will drop by an amount proportional to the hatched area of the dotted curve lying outside D_2. Similarly, as illustrated in Fig. 4.4b, should the pulse amplitude distribution become distorted, the count rate will again drop by an amount proportional to hatched portion of the pulse amplitude distribution.

Shifts and distortions occuring particularly when the flow counter is used fall into four main categories:
  a. Simple pulse amplitude shifts.
  b. Simple pulse amplitude distortions.
  c. Additional pulse amplitude distributions arising from the same energy giving the main pulse amplitude distributions.

d. Additional pulse amplitude distributions arising from a different energy from that giving the main pulse amplitude distribution.

It is the intention to discuss each of these types in turn and to give typical examples.

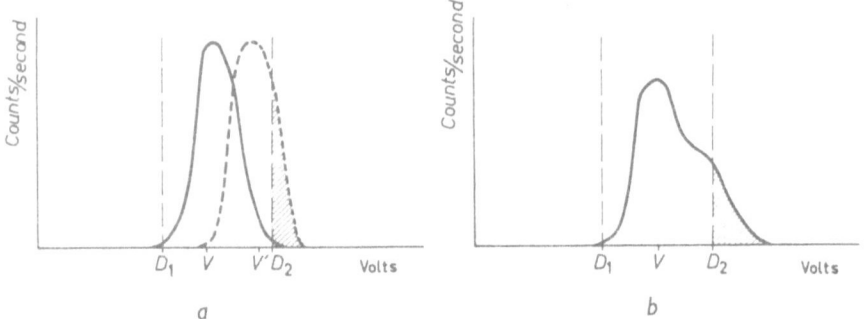

Fig. 4.4b Effect of pulse amplitude shift (a) and pulse amplitude distortion (b) on the peak counting rate of a wavelength corresponding to V. Each effect causes a lowering in count rate by an amount corresponding to the hatched portion of the curve.

4.6 Pulse amplitude shifts

Pulse amplitude shifts are very common and can occur for a variety of reasons. In every case they cause a drop in the measured count rate which can only be restored, either by re-setting the discriminator level, or by altering the total gain of the system.

4.6.1 EFFECT OF COUNTER VOLTAGE

One very obvious possible cause for pulse amplitude shift is instability of counter voltage. As can be seen from Fig. 3.1 gas (or photomultiplier) gain of a counter is very dependent upon the applied voltage. A typical flow counter gave a pulse amplitude increase of 25% for a 1.5% increase in applied voltage. From this, it is apparent that the required stability should be better than 0.1% and in general, the stabilization is such that the equivalent pulse amplitude shift is less than 1% which in fact, normally entails a counter voltage stability of around 0.05%.

4.6.2 COUNT RATE EFFECT

A typical example of a pulse amplitude shift is the effect of count rate illustrated in Fig. 4.5. The upper part of the figure illustrates the normal sequence of events where each individual pulse leaving the counter is completely resolved from its immediate neighbours and the measured pulse amplitude is equivalent to E_A. At high count rates, however, the resolution of individual pulses is not complete, i.e. a pulse starts to grow before the events associated with the previous pulse have terminated. The overall effect is shown in the lower part of the figure. As the pulses never completely decay the practical base line lies at some value above the true zero and the effective pulse amplitude is equivalent to $E_A - E_B$.

This dependence of pulse amplitude on counting rate can lead to serious problems in quantitative X-ray spectrometry since in cases where the pulse height selector is employed the pulses tend to move further and further outside of the selected window setting with increase of count rate. The result is thus rather similar to that of dead time and indeed the two effects are often confused with each other. Although the effect of pulse amplitude dependance upon counting rate has been reported by many workers much of the experimental data is confusing and reported count rate shifts often vary by as much as a factor of two. For this reason it is important to appreciate the reasons for the effect and to evaluate possible methods of overcoming the problem.

It can be shown [27] that for a counter of fixed dimensions operating at a certain voltage and containing a constant gas mixture the only parameter left which could affect the gas gain is the mean free path of the electron. If

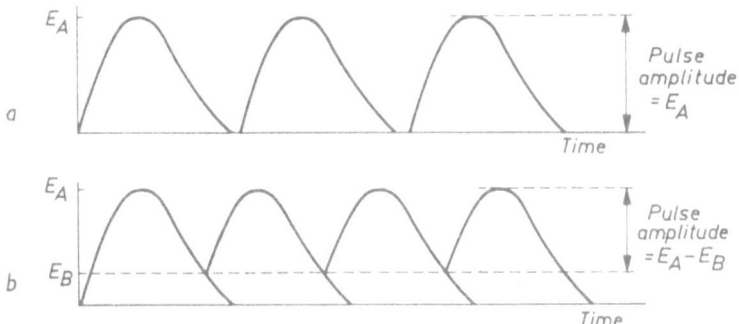

Fig. 4.5 Shift of pulse amplitude with count rate. At low counting rates (upper diagram) each individual pulse is completely resolved from its neighbours and the pulse amplitude is proportional to E_A. At high counting rates (lower diagram) a new pulse starts to grow before the events associated with the decay of the previous pulse have terminated, with the result that an artificial zero is set up at E_B and the measured pulse amplitude is given by $E_A - E_B$.

this is indeed the case it would be expected that positive ion density be the critical factor, since under constant field conditions (i.e. fixed counter dimensions and fixed anode voltage) this is the only thing that could affect the mean free path of the electron.

Experimental results confirm this hypothesis the following factors having been found to influence the degree of pulse amplitude shift:

a. Counting rate (which determines the number of photons entering the counter)

b. Energy of the measured radiation (which determines the number of primary ion pairs per incident photon)

c. Voltage on the anode (on which the gas gain depends)

d. Dirt on counter anode (which causes variations in the field around the anode)

e. Anode diameter (on which the ion density depends)

Increase of any of the first four mentioned factors will in turn increase the degree of pulse amplitude shift. Increasing the diameter of the anode increases the radius of the ionization zone, thus at a fixed gas gain the same number of positive ions are contained within a larger ionization volume thus the effective ion density is lower. Thus a thicker anode (of similar smoothness and cleanliness) will always give smaller pulse amplitude shifts. From the above it will be obvious that count rate/pulse amplitude shift data rae of little

TABLE 4.1

Summary of the effects producing variable pulse shift

Variable	Anode diameter (in μm)	Voltage	Counter resolution for FeKα (%)	Incident radiation	Pulse shift(%) 10^3–5×10^4 c/s
Effect of dirt	25/100 clean	(1)	13.6	SKα	10.1
		(1)	17.1	SKα	8.7
	25/100 dirty	(1)	22.1	SKα	30.5
		(1)	19.3	SKα	10.5
Anode diameter	25	1800	15.0	SKα	24.0
	50	1930	15.2	SKα	21.5
	100	2300	17.8	SKα	11.0
	150	2660	21.7	SKα	8.7
Counter voltage	25	1400	(2)	SKα	3.5
	25	1600	(2)	SKα	10.1
	25	1800	(2)	SKα	24.8
Energy of incident radiation	25	(1)	(2)	SKα	24.0
	25	(1)	(2)	KKα	32.7
	25	(1)	(2)	FeKα	36.2

Note: (1) Fixed gas gain
(2) Average counter cleanliness (i.e. about 15% for Fe Kα)

value unless certain experimental conditions are quoted including energy of
the measured radiation, voltage on the counter (or better, the gas gain) the
anode (and counter) diameter and the cleanliness of the counter. The best
means of establishing the latter point is by study of the detector resolution
value for a radiation of a certain energy.

Table 4.1 summarizes the various contributions to the pulse amplitude
shift effect. As will be seen an attempt has been made to establish the optimum anode diameter by use of a wide range of anode thickness between 25
and 150 μm. At a fixed gas gain an increase of anode diameter decreases
the pulse amplitude shift and what is more important, reduces the relative
effect of dirt particles. The optimum appears to lie around 100 μm since
thicker anodes tend to give poor detector resolution-presumably due to the
formation of a distorted or asymetric volume of ionization around the
anode. Also the voltage requirement increases rapidly with the anode diameter.

Count rate shift problems are thus minimized by using the lowest possible
gas gain that can be tolerated without excessive noise problems (this in effect
means the lowest voltage possible), using an anode diameter of about 100
μm and keeping the inside of the counter — particularly the anode wire —
scrupulously clean.

4.6.3 GAS DENSITY EFFECT

A far less easily overcome problem is that of temperature and pressure
effects on the flow counter gas. Since the magnitude of the gas multiplication
process is dependent upon the kinetic energy gained by the ionised electron
in its mean free path, anything which causes the length of the mean free path
to change will also vary the amplitude of the pulse leaving the counter. As
the mean free path is dependent upon the average distance between gas
atoms, the total number of atoms contained within the volume of the counter
will be a critical factor. As the counter is necessarily of fixed dimensions the
number of atoms contained within it will be directly proportional to the
density of the gas and this in turn will be dependent upon gas temperature
and pressure. In general, an increase in gas density decreases the electron
mean free path and normally decreases the pulse amplitude; hence an
increase in pressure causes a lowering of the pulse amplitude and an increase
in temperature causes the pulse amplitude to increase. The magnitude of the
effect depends to a large extent on the design of the counter, but data from a
typical counter showed that a 1 % change in pressure or absolute temperature
brought about a 6 % shift in the pulse amplitude.

Much can be done to minimise the overall effect on the measured count rate for any particular wavelength by careful setting of the pulse height selector, making due allowance for probable changes in laboratory temperature and pressure. However, such a procedure can only be adopted at the expense of a lower pulse height selector efficiency because of the excessively wide window required to embrace the likely shifts in the pulse amplitude distribution. There is little incentive for temperature stabilisation alone, since the pressure effect is of equal, if not greater, importance.

It is apparent that a gas density compensator is required and one such device which has recently been developed is that fitted to the Philips PW series automatic X-ray spectrometers. The device is of fairly simple construction and a simplified drawing is shown in Fig. 4.6. It consists of a cylinder of fixed volume which is fitted with an inlet port and an outlet port. A bellows device is fixed at the top of the cylinder with an adjustable clamp and the bottom of the bellows is free to move up and down with volume changes of the gas inside the bellows. The lower part of the bellows is attached to a lever arm which is pivoted at one end and which supports a conical needle at the other. This needle restricts the flow of incoming gas by an amount dependent upon the position of the needle arm which, in turn, is determined by the volume of the gas inside the bellows. In operation, the assembly is placed in the flow gas system between the reducing valve and the flow counter inlet and the effluent gas leaving the flow counter assembly is discharged to the atmosphere via a fixed orifice. The principle of operation depends upon keeping the pressure inside the bellows the same as the pressure inside the cylinder and this is done by automatic variance of the flow rate through the

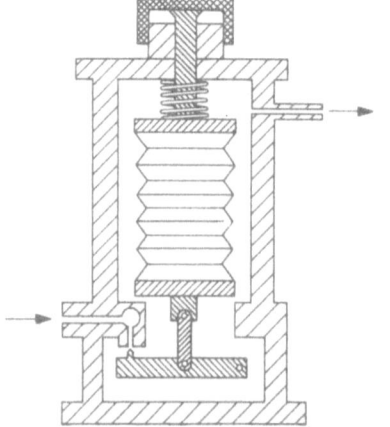

Fig. 4.6 The gas density compensator. The effect of temperature and pressure on the fixed volume of gas inside the bellows, is used to vary the flow of gas through the system by actuating a needle valve in the inlet line. The subsequent variation in the build up of pressure against a fixed orifice serves to compensate for the changes in the number of gas atoms in the fixed volume of the counter body, caused by changes in temperature and pressure.

fixed orifice which in turn varies the build-up of pressure against the needle valve. This particular density compensating valve has been found to be effective over the normal range of laboratory temperatures and pressures.

4.6.4 EFFECT OF IONISABLE GAS ATOMS TO QUENCH GAS RATIO

Another effect causing pulse amplitude shift is found when the bottle supplying the argon-methane gas mixture to the counter is almost empty. [28]

As there is a large difference in the partial pressures of argon and methane a significant change in composition is found when samples of the supply gas from a full bottle are compared with similar samples taken from an almost empty bottle. Mass spectrometric analysis of typical samples from almost empty and full cylinders show changes of the order of 0.5 % relative in the methane content which is nominally 10 %. Due to the mechanism of the gas amplification process the ratio of ionisable gas atoms to quench gas atoms is known to have a marked effect on the pulse amplitude, [29] and the equivalent pulse amplitude shift corresponding to the change in methane content reported above, was found to be of the order of 10 %. In general, this effect does not manifest itself unless the gas supply bottle is more than 95 % discharged and problems can be avoided by rejecting supply gas bottles whilst about 5 % of their contents are still available. In addition, it is good practice to check the pulse amplitude of a typical and standard wavelength each time a supply gas bottle is changed.

4.7 Pulse amplitude distortions

Since the majority of the gas amplification occurs within a few wire diameters of the anode, it is important that a symmetrical and homogeneous field be maintained within this active volume. [30-32] Any distortion of the field can cause an equivalent distortion in the pulse amplitude distribution and in the worst case may give more than one peak maximum.

End effects which arise from the clamping of the anode wire [31] are normally avoided by use of an axial window but problems may still arise from irregularities in the diameter of the anode wire or from the deposition of dirt carried into the body of the counter with the incoming flow gas. It is common practice to clean fresh anode wire before mounting in the counter but unless special precautions are taken to prevent the intake of dirt the resolution of the counter can rapidly deteriorate to up to twice its theoretical value. For example, a recent survey amongst ten users of identical automatic spectrometers, in which no special precautions had been taken to preclude dirt, showed that the resolution for FeKα radiation (theoretical resolution about 17 %) varied from 17.6 % for the newest installation (6 weeks of use) to 35.7 % for the oldest installation (about 2 years of use) and in addition, a rough correlation could be shown between the measured counter resolution and the period of use.

Fig. 4.7a shows a typically distorted pulse amplitude distribution which has been measured with a dirty flow counter using titanium Kα radiation. The measured resolution of 28.3 % is almost 70 % worse than the originally measured value of 18.6 %. Following the careful cleaning of the anode wire using a camel hair brush soaked in benzene, a greatly improved resolution was obtained, (see fig. 4.7b), and a value of 19 % was obtained for the resolution. Pitting of the anode wire may follow the combined action of the dirt and traces of oxygen in the flow gas and, should this occur, the only remedy is to replace the anode wire.

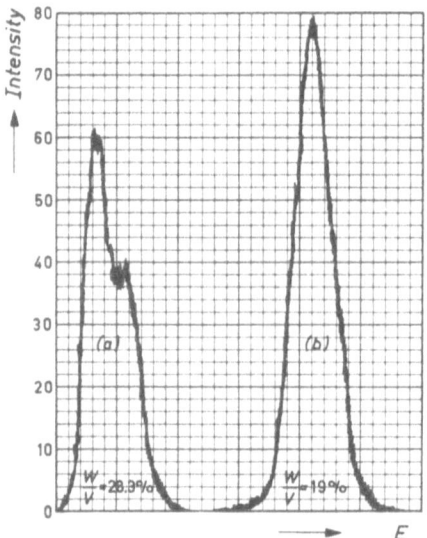

Fig. 4.7 Effect of dirt on the resolution of the flow counter. Particles of dirt which may be carried into the body of the flow counter by the effluent gas, lodge on the anode wire distorting the field. The resulting distortion of the pulse amplitude distribution is shown in figure (a). By removing the dirt from the anode the symmetry of the field is restored and the counter resolution is again normal, figure (b).

4.7.1 ADDITIONAL PEAKS ARISING FROM THE SAME WAVELENGTH GIVING MAIN PEAK

If the X-ray photons to be measured produce more than one pulse amplitude distribution, the possiblities of the use of either or both the pulse amplitude distributions have to be considered. Such a circumstance arises when an escape peak occurs along with the natural peak and, in general, the practicability of increasing the window width to embrace both escape and natural peaks depends upon the relative amplitudes of the two peaks. For elements which give rise to escape and natural peaks, if the count rate due to the escape peak is measured at the same time as that due to the natural peak, the total count rate arising from the measured wavelength can be increased by as much as 5-10 % for an argon-methane flow counter. However, there is little incentive for opening up the window to embrace an escape peak in addition to the natural peak if, by doing so, the background count rate is increased by an equivalent amount. Such a situation might well arise if the escape and natural peaks lie on an increasing background as described earlier with reference to Fig. 4.3. Table 4.2 shows the relative positions of escape and natural peaks arising from the interaction of various $K\alpha$ radiations with argon. It will be seen that as the energy of the incident radiation increases, so the relative separation of the natural and escape peaks decreases. For

TABLE 4.2

Relative separations of escape and natural peaks arising from the same wavelength

Element	Energy (E)	$E - E_a$	Relative position of escape peak assuming natural peak position of 10 units
K	3.31	0.36	1.09
Ca	3.69	0.74	2.01
Sc	4.09	1.14	2.79
Ti	4.51	1.56	3.46
V	4.95	2.00	4.04
Cr	5.41	2.46	4.54
Mn	5.90	2.95	4.99
Fe	6.40	3.45	5.39
Co	6.92	3.97	5.73
Ni	7.47	4.52	6.07

The relative amplitude of the escape peak (V_e) is given by

$$\frac{V_e}{V_n} = \frac{E - E_a}{E}$$

where V_n is the amplitude of the natural peak arising from an energy E, and E_a is the energy of $ArK\alpha$.

normal settings of the pulse height seléctor, the escape peak arising from potassium radiation is lost in electronic noise and is not observed. In addition, the escape and natural peaks are not usually resolved above atomic number 27 and hence only a fairly narrow range of elements from scandium ($Z = 21$) to cobalt ($Z = 27$) are likely to present a problem.

A corresponding situation will exist for each type of gas filling and the magnitude of the effect will depend upon the fluorescent yield of the counter gas. For example, as illustrated in Fig. 4.8 neon has a very low K fluorescent yield (about 0.08) and gives no measurable escape peaks,[21] krypton however, has a high K fluorescent yield (about 0.51) and gives a large escape peak. Xenon with a K fluorescent yield of 0.71 has an escape peak three times the intensity of the natural peak. (The pulse amplitude maximum at 9.5 keV is due to CeL radiation passing over the top of the crystal into the detector). Excape peaks can also occur following the interaction of L or M series characteristic radiation with the counter gas. However, since these radiations invariably fall within a longer wavelength range than the K series the fluorescent yield values are in general lower and equivalent escape peak intensities are similarly low.

In the case of the scintillation counter escape peak phenomena can also occur following the ejection of iodine K or L electrons, but as the absorption edge energies of iodine fall outside the usual operating range of the scintillation counter, escape peaks are rarely observed. The iodine L absorption edge value of 5 keV is near the accepted long wavelength limit of the scintillation counter and unless K wavelengths, shorter than 0.37 Å i.e. greater than atomic number 57 (lanthanum), are being measured the iodine K absorption edge value of 33 keV is not exceeded.

An interesting phenomenum can sometimes be observed when automatic pulse height selection is applied in the recording of the spectrum of a low average atomic number sample where the background is high owing to scatter.[33] If a scintillation counter is used as the detector and a topaz crystal is used to disperse the radiation a discontinuity occurs in the background at about 15.8°2 θ. This angle corresponds to an energy of 0.37 Å which is equivalent to the iodine K absorption edge value. The reason for the discontinuity is that below 0.37 Å (less than 15.8°2 θ) the continuous radiation is sufficiently energetic to excite iodine Kα radiation from the sodium iodide in the scintillation counter phosphor. Hence an escape peak is formed in addition to the natural peak but the amplitude of the escape peak falls outside the window of the pulse height selector which is set to receive the natural pulse amplitude. As a result the background count rate is lower than would be anticipated. Above 0.37 Å (greater than 15.8° 2 θ) the white radiation is insufficiently energetic to excite iodine Kα radiation and any pulses which arise from the radiation absorbed in the phosphor occur at an amplitude corresponding to the measured radiation, resulting in a normal background count rate. Another extra peak which may be observed when recording the complete pulse height distribution at high counting rates is

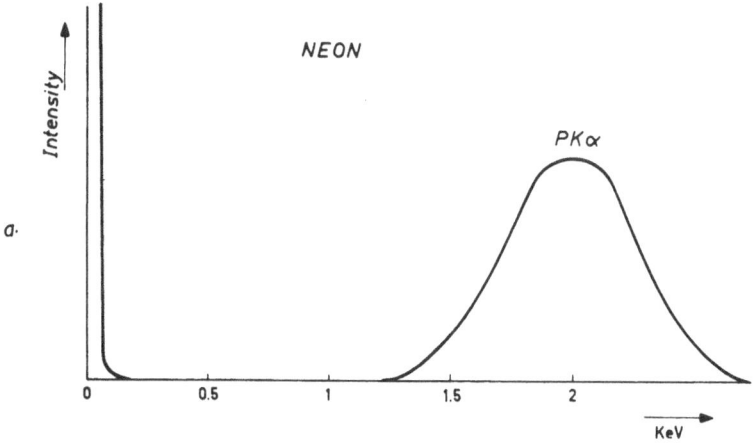

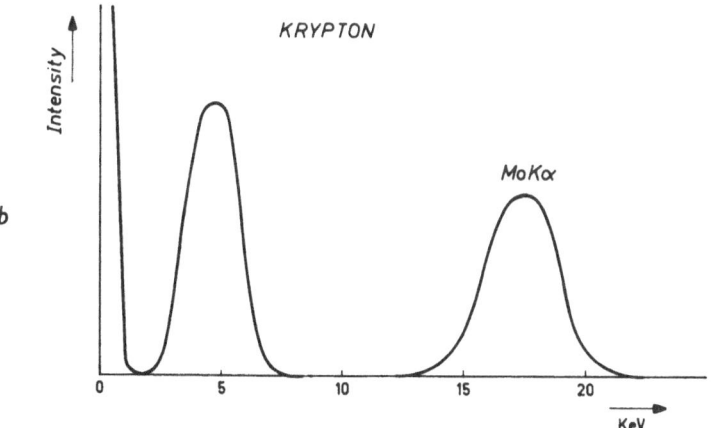

Fig. 4.8 Pulse amplitude distributions from different counter gases plotted in each instance for a wavelength just shorter than that of the appropriate counter gas.

the double or sum peak. If a photon enters the detector whilst a previously entered photon is still being counted, the result may be a pulse of approximately twice the amplitude, occurring in the pulse height diagram at double the voltage of the main peak. If the pulse height selector is set to register the main peak, then twice the number of "doubled" pulses are lost, leading to apparently higher detector dead time.

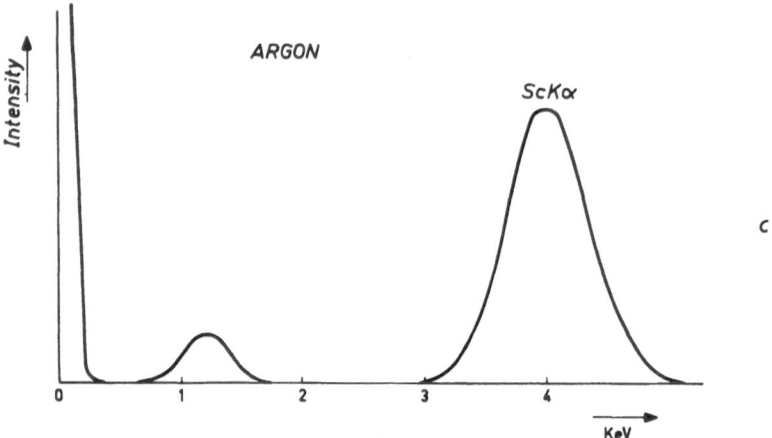

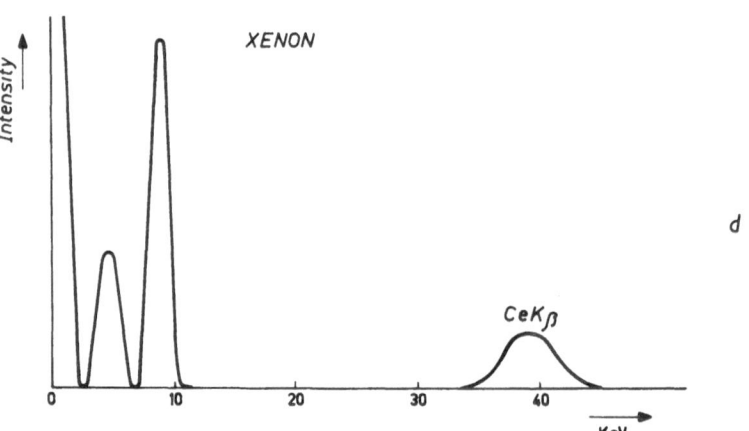

4.7.2 ADDITIONAL PEAKS NOT ARISING FROM THE MEASURED WAVELENGTH

Circumstances can occasionally arise where additional pulse amplitude distributions occur within the operating range of the pulse height selector and where these extra distributions are not directly associated with the measured radiation [34]. A common example of this is the effect due to additional wavelengths (other than harmonic overlapping wavelengths) entering the detector which has been set nominally to measure one specific wavelength. This effect is due to the finite solid angle of acceptance of the detector and the result is typical of the situation shown in Fig. 4.2. When phosphorus $K\alpha$ is being measured in the presence of high concentrations of calcium using a P.E. or E.D.D.T. crystal, if primary collimation of greater than 400 microns spacing is employed, some of the calcium radiation enters the detector due to partial harmonic overlap phosphorus $K\alpha$ and calcium $2K\beta$. The net effect is that three peaks are observed in the pulse amplitude distribution which in Fig. 4.2 would correspond to V_2 being phosphorus $K\alpha$, V_3 the calcium $2K\beta$ and V_1 the escape peak associated with the calcium $2K\beta$. The count rate due to calcium can of course be completely removed by setting up the pulse height selector with a lower level at D_1 and a window width of $(D_2 - D_1)$, or alternatively by using finer collimation such that the calcium and phosphorus lines are completely resolved before entering the detector.

A second effect falling within this general category is that due to crystal fluorescence. If the analysing crystal contains elements which can give rise to characteristic radiations which fall within the detection range of the counter, additional peaks will again occur in the general pulse amplitude distribution. Fig. 4.9 demonstrates a typical case which occurs when sodium $K\alpha$ radiation is measured making use of a gypsum crystal. In addition to a peak for sodium $K\alpha$, peaks also occur corresponding to the K radiations of sulphur and calcium.

Crystal fluorescence can sometimes be the cause of an apparent count rate shift in the pulse amplitude. If, for example, phosphorus is being determined under standardized conditions, using a gypsum crystal, the measured pulse

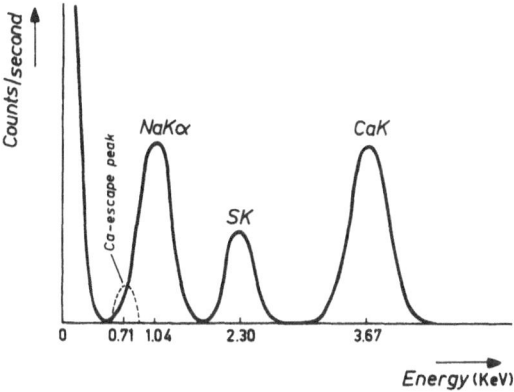

Fig. 4.9 Effect of crystal fluorescence. When a gypsum (calcium sulphate) crystal is used as a dispersing crystal for the analysis of, for example, sodium, the incident radiation may excite both calcium and sulphur in the crystal matrix. As the K radiations of both of these elements fall within the energy range of the pulse height selector, peaks will occur corresponding to both calcium and sulphur as well as for sodium.

amplitude is found to shift to a lower value as the phosphorus concentration in the sample is increased. This effect arises from the close proximity of the pulse amplitudes of phosphorus $K\alpha$ radiation from the sample with that due to sulphur K radiation from crystal fluorescence. Although these amplitudes differ by about 12%, they are not resolved, hence the measured pulse amplitude is a composite of the two. At low phosphorus concentrations the pulse amplitude due to sulphur is the limiting factor in the composite distribution which therefore peaks at the position of sulphur K radiation. However, as the phosphorus concentration is increased the effect of its pulse amplitude distribution on the composite distribution becomes more and more significant until eventually it peaks at an amplitude corresponding to phosphorus $K\alpha$ radiation.

In general, provided that the flow counter has sufficient resolution to resolve the extra distributions from that due to the required wavelength, the effect of crystal fluorescence on the analytical procedure will not be important. However, if the extra lines are not resolved, a significant count rate will occur at all wavelength thus at the wavelength of the required element and a significant intensity will occur even though this element is not present. For example, in the determination of phosphorus, sulphur, chlorine, potassium, calcium and scandium using a gypsum crystal, these difficulties do arise. Other commonly used crystals which show this effect include potassium acid phthalate, sodium chloride and ammonium dihydrogen phosphate.

REFERENCES

1   FRANCIS, J. E., BELL, P. R. and GUNOLACH, J. C., 1951, Rev. Sci. Instr., **22**, 133.
2   VAN RENNES, A. B., 1952, **10**, 20, Nucleonics No. 7.
3   VAN RENNES, A. B., 1952, **10**, 22, Nucleonics No. 8.
4   MILLER, D. C., 1957, Norelco Reporter, **4**, 37.
*   5   HEINRICH, *Advances in X-Ray Analysis*, Plenum, New York, 1961, **4**, 370.
6   MALMSTADT, ENKE and TOREN, *Electronics for Scientists*, Benjamin, New York, 1963.
7   LANG, A. R., G.P. 1023246/1954.
8   WYTZES, S. A., Philips Technical Review, **27**,11.
9   BEHN RIGGS, F., 1963, Rev. Sci. Instr., **34**,312.
10 MARCHAL, J. and WEBER, K., 1964, J. Sci. Instr., **41**, 15.
*   11 PARRISH, W. and KOHLER, T. R., 1956, Rev. Sci. Instr., **27**, 795.
12 MAEDER, D., 1958, Nuclear Instrum., **2**, 324.
13 BISI, A. and ZAPPA, L., 1955, Nuovo cim., **2**, 298.
*   14 CURRAN, S. C., COCKROFT, A. L. and ANGUS, J., 1949, Phil. Mag., **40**, 929.
*   15 HENDEE, C. H. and FINE, S., 1954, Phys. Rev., **95**, 281.
*   16 LANG, A. R., 1952, Proc. Phys. Soc., **65**, 372.
*   17 MULVEY, T. and CAMPBELL, A. J., 1958, Brit. J. Appl. Phys., **9**, 406.
18 KOHLER, T. R. and PARRISH, W., 1955, Rev. Sci. Instr., **26**, 374.
19 TANEMURA, T., 1961, Rev. Sci. Instr., **32**, 364.
*   20 SEIBEL, G., 1964, J. Appl. Radiation and Isotopes, **15**, 25.
*   21 JENKINS, R., 1965, Rev. Sci. Instr., **42**, 480.

22　HALL, *Advances in X-Ray Analysis*, Plenum, New York, 1957, **1**, 297.
23　BESSEN, *Advances in X-Ray Analysis*, Plenum, New York, 1957, **1**, 455.
24　DOLBY, R. M., 1959, Proc. Phys. Soc., **73**, 81.
25　BERTOLINI, G., BISI, A. and ZAPPA, L., 1953, Nuovo cim., **10**, 1424.
26　BIRKS, L. S. and BATT, A. P., 1963, Analyt. Chem., **35**, 778.
27　JENKINS, R., 1968, Philips Scientific Reports, 79,136. FS6, Philips, Eindhoven.
28　JENKINS, R., 1964, Rev. Sci. Instr., **41**, 696.
29　CURRAN and CRAGGS, *Counting Tubes*, Butterworths, London, 1949.
30　SHARPE, *Nuclear Radiation Detectors*, Methuen, London, 1964.
31　COCKROFT, A. L. and CURRAN, S. C., 1951, Rev. Sci. Instr., **22**, 37.
32　LONG, A., 1959, J. Brit. I.R.E., **19**, 273.
33　PARRISH, *Advances in X-Ray Analysis*, Plenum, New York, 1965, **8**, 118.
34　JENKINS, R., and HURLEY, P. W., 1968, Canadian Spectroscopy, **13**, 35.

CHAPTER 5

COUNTING STATISTICS

5.1 Introduction

The net intensity of emitted characteristic X-radiation from an element in a matrix is related to the concentration of that element. Fig. 5.1 illustrates the theoretical correlation between the peak intensity R_p of an element with its concentration C. The true background response R_b is given by the intercept of the curve on the ordinate. The slope m of the curve, sometimes called

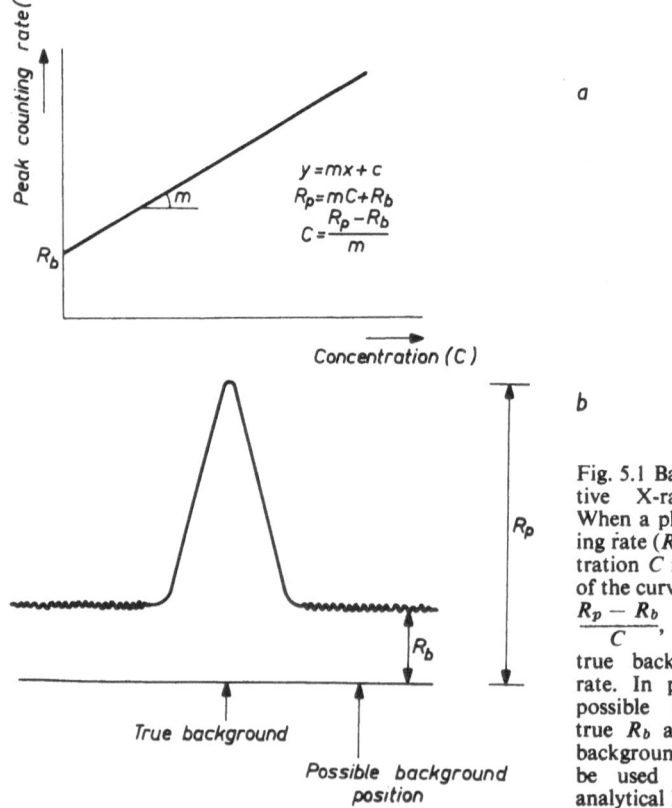

$$y = mx + c$$
$$R_p = mC + R_b$$
$$C = \frac{R_p - R_b}{m}$$

Fig. 5.1 Basis of quantitative X-ray spectrometry. When a plot of peak counting rate (R_p) against concentration C is made, the slope of the curve is equal to $\frac{R_p - R_b}{C}$, where R_b is the true background counting rate. In practice it is not possible to measure the true R_b and an equivalent background position has to be used away from the analytical line.

the calibration factor, is equal to $\dfrac{R_p - R_b}{C}$ and the concentration is given by

$$C = \frac{R_p - R_b}{m} \tag{5.1}$$

By use of standards the value of m, for a certain element in a particular type of matrix, can be determined and any unknown concentration falling within the calibration range can be determined by substituting measured values of R_p and R_b in equation (5.1). Theoretically the net intensity ($R_p - R_b$) is equal to the peak intensity of a wavelength from the element in the matrix, minus the true background intensity i.e. the intensity measured at the same wavelength, using identical conditions as for the peak measurement, in an identical matrix, but where the analysed element is absent. In practice the true background position can never be measured since the mere fact that the required element is necessarily absent from the "blank" matrix means that the characteristics of the "true blank" matrix and the "actual blank" matrix are different. The overall accuracy in the determination of the concentration of an element in a matrix is dependant upon the accuracy with which both peak and background responses can be determined, as well as upon the reliability of the calibration factor. Each of these parameters is subject to certain random and systematic errors which to a first approximation define the precision and the accuracy of the determination respectively. The errors arise from a number of sources of which the most important (but by no means the only) are limited apparatus reproducibility, matrix interferences and the statistical nature of emission and detection of X-rays. This particular chapter will deal exclusively with the third of these sources of error.

5.2 Definition of statistical terms

Much confusion has arisen in the minds of many spectroscopists owing to the apparently numerous methods of reporting data and quoting the reliability of their analyses. There have been several attempts by various interested bodies over recent years to clarify this situation and the following list of more important definitions has been compiled from the reports of these bodies.[1-4] It is not the purpose of this book, nor indeed the intention of the authors, to introduce a mass of statistical data and complex mathematical proofs pertaining to the determination of precision and reliability of measurements involving random fluctuations. What will be done, however, is to provide working equations and examples of their use in areas which the reader is likely to cover in the application of X-ray spectrometry. For further information on the statistical handling of data the reader is referred to suitable works covering this topic.[5-7]

True Result: The true result is that concentration which actually exists in a representative sample i.e. the right concentration. This may never be known absolutely but in a standard sample, a chosen concentration sometimes called the *accepted reference level*, defined as the most accurately known or agreed value, is often used as the true result.

Mean (Average, Arithmetic Mean, $\bar{x}$): is the arithmetic sum of a set of results $x_i — x_n$ divided by the number (n) of results.

$$\bar{x} = \frac{1}{n} \sum_{i=1}^{i=n} x_i$$

Median: This is the middle value of a series of results placed in ascending order. Where there is an even number of results, the arithmetic mean of the middle pair of v.lues is taken.

Precision: The closeness of agreement among replicate results obtained under any definite set of conditions, usually expressed as a relative error. An increase in precision implies a decrease in its absolute numerical value.

Random Error: Random errors arise from fluctuations in experimental conditions and the inherent errors in the method of measurement. If is found in practice that the random errors which arise from variations in most natural processes approximate to a *Normal* or *Gaussian* distribution,

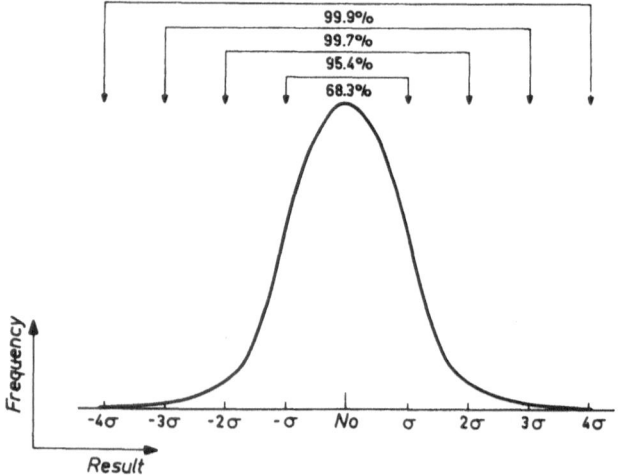

Fig. 5.2 The Gaussian distribution. Provided that the number of measurements is large there is a 68.3% probability that a single measurement will lie between $N_0 \pm \sigma$, 95.4% that it will lie between $N_0 \pm 2\sigma$, 99.7% that it will lie between $N_0 \pm 3\sigma$ and 99.9% that it will lie between $N_0 \pm 4\sigma$.

(see Fig. 5.2) hence the significance of any error can be defined in terms of the parameters of this distribution.

Accuracy: Can be regarded as the nearness of a result or mean of a number of results to the true result. The accuracy of a result is therefore dependant both upon the precision of the measurement and on the difference between the true result and the experimental result. This difference is defined as the *Bias* or *Systematic Error*. Therefore, a method or measurement may be precise without being accurate.

Standard Deviation: Standard deviation is defined as the root-mean-square deviation of a set of observations from their arithmetic mean. It is customary to designate the standard deviation as σ for an infinite number of measurements with a mean of μ, or as S for a finite number (n) of measurements with a mean of $\bar{x}$. Thus

$$\sigma = \sqrt{\frac{\Sigma(x - \mu)^2}{n}} \quad \text{and} \quad S = \sqrt{\frac{\Sigma(x - \bar{x})^2}{n - 1}}$$

Variance (V): The variance is the square of the standard deviation.

$$V = S^2$$

Coefficient of Variation (C.V.) (alternatively, relative standard deviation ε): is defined as the standard deviation expressed as a percentage of the mean.

$$C.V. = \frac{100\,S}{\bar{x}} \,\%$$

Confidence Limits: The confidence limits are the upper and lower values between which the actual measurement will fall with a certain probability.

For an infinitely large number of readings showing a normal distribution there is a 68.3 % probability that a single result will fall within $\pm 1\,\sigma$ of the mean of the distribution, a 95.4 % probability that it will fall within $\pm 2\,\sigma$ and a 99.7 % probability that it will fall within $\pm 3\,\sigma$.

Probable Fractional Error: Where there is a 50 % probability of a single result falling within $\pm 0.67\,\sigma$ of the mean, the situation is referred to as the probable fractional error and is designated ε_{50}. ε_{90} similarly refers to $1.64\,\sigma$ and ε_{99} to $2.58\,\sigma$.

Repeatability:*) This usually refers to single-laboratory, multi-operator-machine-day precision.

Reproducibility:*) This usually refers to multi-laboratory-machine-operator-day precision.

---

*) Other combinations of laboratory-operator-machine-day are sometimes used in the definition of these terms, hence care must be taken in their use.

Tolerance Limits: The limiting values between which measurements must lie if a result is to be acceptable, as distinct from confidence limits.

Flyers, Outliers, Rogue values, Wild values: In general these are results which deviate markedly from the other members of the population. They may be due either to an extreme manifestation of the random variability inherent in the data, or to mistakes such as typographical errors, wrongly identified samples, mis-setting of the instrument and so on.

5.3 Random distribution of X-rays

If an X-ray measurement consisting of the determination of a number of counts N was repeated many times it would be found that the values obtained fall within a definite distribution about the true value N_0 and provided that the number of measurements was large the distribution would approximate to a Gaussian distribution. In point of fact the random distribution of X-rays always follows a Poisson distribution

$$W(N) = \frac{(N_0)^N}{N!} \exp[-N_0] \tag{5.2}$$

which in turn approximates to a Gaussian distribution

$$W(N) = \frac{1}{\sqrt{2\pi N}} \cdot \exp\left| -\frac{(N - N_0)^2}{2N} \right| \tag{5.3}$$

provided that N is large i.e. $\sqrt{N} >> 1$. The standard deviation (σ) of distribution is equal to $\sqrt{N_0}$ which is in turn approximately equal to $\sqrt{N}$. From the properties of a Gaussian distribution there will be a 68.3% probability that any value of N will be between $N_0 \pm \sigma$. It is important to keep in mind that a 95.4% probability exists for any value N to be within an interval $N_0 \pm 2\sigma$ and that a 99.7% probability exists that any result N will fall into an interval $N_0 \pm 3\sigma$. If, for example, N is 10,000, there is a 95.4% probability that any measurement will fall between 9,800 and 10,200 $(10,000 \pm 2 \cdot \sqrt{10,000})$.

In general N is known rather than N_0, however, since the distribution is Gaussian and $\sqrt{N}$ differs little from $\sqrt{N_0}$ it can be assumed that N_0 will be in an interval $\pm 2\sigma$ around the measured value N in 95.4% of the cases. In this way information can be obtained on the probable value of N_0. For example if the measured value of $N = 9900$, it could be concluded that in 95.4% of the cases N_0 lies between $N \pm 2\sigma$ of N i.e. between 9900 $\pm 2 \cdot \sqrt{9,900}$ or 9,700 and 10,100.

The fact that N fluctuates around N_0 makes it possible to check whether these fluctuations are normal (statistical). In practice the average $(\overline{N})$ is considered a fair approximation of N_0. In general if many measured values N differ more than 3 σ from $\overline{N}$, other souces of deviation are present.

It will be appreciated that one is actually interested in the standard deviation of a counting rate (R) rather than a number of counts (N) but since

$$N = RT \tag{5.4}$$

where T is the counting time, provided that the error associated with the measurement of T is insignificant and does not exihibit a random distribution, then:

$$\frac{\sigma(R)}{R} = \frac{\sigma(N)}{N} \tag{5.5}$$

It should be remembered, however, that equation (5.5) is only an approximation and where high precision is required in short analysis times the true significance of the error in the measurement of T should be established.

The fact that the standard deviation of any measurement can be related to the number of quanta measured is useful since provided that the error due to counting statistics is the limiting error, the precision of the measurement can be predicted. Furthermore, provided that systematic errors such as those arising from matrix effects are negligible, the final accuracy of the measurement can also be predicted simply from the number of counts taken.

ε is obtained by relating the absolute standard deviation σ_N to N or

$$\varepsilon = \frac{\sqrt{N}}{N} = \frac{1}{\sqrt{N}} = \frac{1}{\sqrt{RT}} \tag{5.6}$$

or in percentages

$$\varepsilon \text{ or } \sigma \% = \frac{100}{\sqrt{N}} \text{ or } \frac{100}{\sqrt{RT}} \tag{5.7}$$

The standard deviation σ_R on the count rate R is

$$\sigma_R = \varepsilon \cdot R = \sqrt{\frac{R}{T}} \tag{5.8}$$

The following data shows that the longer the counting time the lower ε, provided the equipment is stable during that time.

Variation of ε with time, -given that $R = 100$ cps

T (s)	1	4	10	40	100	400
$\varepsilon \%$	10	5	3.17	1.58	1	0.5

Alternatively one can say that $\sigma \%$ (usually called coefficient of variation or

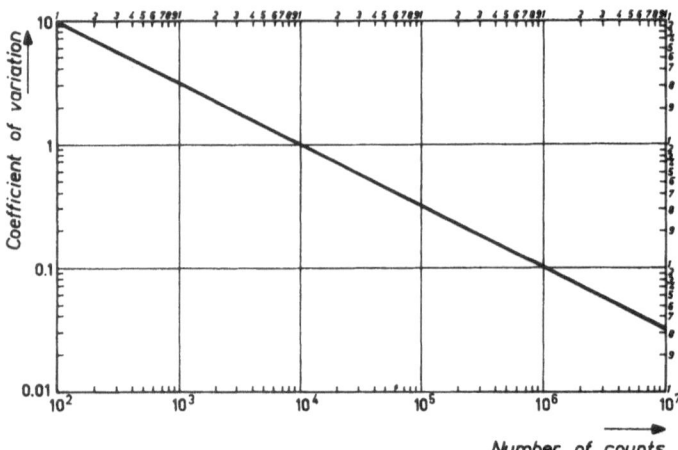

Fig. 5.3 Correlation of the coefficient of variation and the number of counts taken.

relative standard deviation) decreases with increase of N in the manner shown in Fig. 5.3.

5.4 Choice of fixed time or fixed count

In the simple case where background count rates can be ignored it will be seen from Equation (5.4) that there are two methods of determining the peak counting rate — either by measuring the time required to collect a fixed number of counts (fixed count method F.C.), or by measuring the number of counts collected in a selected time (fixed time method F.T.). The particular choice depends to a large extent upon circumstances but in general the fixed time method is the more convenient. In addition to any statistical considerations there are two distinct advantages of the fixed time method and the first of these is a purely practical consideration. If a method has been developed for the determination of a certain element in a particular type of sample matrix and the element falls within a certain range, say 15-20%, provided that the element is indeed present at this level the fixed time or fixed count method might be equally convenient. If, however, a sample should be submitted for analysis in which the element is present at a very low level, if fixed counts are being accumulated much time will be lost in collecting the preselected number of counts to determine a concentration which may be outside the calibration range anyway. This factor is particularly important in the programming of automatic X-ray spectrometers.

The second advantage of fixed time over fixed count is illustrated in

Fig. 5.4. If it is assumed that a calibration curve has been prepared for a certain element over the range 0-1 % and, ignoring background, then the slope of the curve is equal to 100 c/s %. If it is decided that a precision (σ) of 1 % is required at the 1 % level, i.e. 1 $\pm$ 0.01 % (σ), it is useful to study the results of fixed count and fixed time methods at the lower end of the range. In the case of the fixed count method an analysis time of 1000s would be required at the 0.1 % level to give a precision of 0.1 $\pm$ 0.001 % (σ). However in the case of the fixed time method $N = 10^3$ and the analysis time of 100s would yield 0.1 $\pm$ 0.003 % (σ) at the same lower level. It is generally true that the analyst would prefer to sacrifice relative precision at the lower end of the concentration range rather than take ten times longer to give him a precision which, relatively speaking, is ten times better than that at the 1 % level.

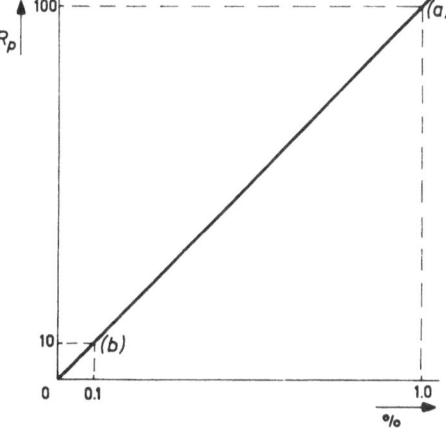

Fig. 5.4 Choice between fixed time and fixed count.

		Fixed Count	Fixed Time
	%	1	1
	N	10,000	10,000
Sample a(1%)	R	100	100
	T	100	100
	Result	1 $\pm$ 0.01 (100 s)	1 $\pm$ 0.01 (100 s)
	N	10000	1000
Sample b(0.1%)	R	10	10
	T	1000	100
	Result	0.1 $\pm$ 0.001 (1000 s)	0.1 $\pm$ 0.003 (100 s)

An advantage of fixed count measurements is that if dead time corrections are required, as indeed they often are, when absorption corrections are being applied, the dead time correction is made very simply by dividing the number of counts collected by the analysis time minus an increment for dead time loss. For instance with an equipment dead time of 3 μs this increment is 3s, where N is 10^6.

Number of Counts	Increment to be subtracted from the counting time
10^6	3 s
10^5	0.3 s
10^4	0.03 s
10^3	0.003 s

Thus if 10^6 counts were collected in 23 seconds the true count rate is equal to $\left(\dfrac{10^6}{23-3} \right)$ c/s $= 50,000$ c/s. Where fixed time methods are employed the calculation is slightly more complicated.

In practice, the situation is invariably more complicated than this since often background cannot be ignored and as will be seen later, this offers a third method of determining R, or rather $(R_p - R_b)$, this being the method of fixed time optimal (F.T.O.).

5.5 Limit of counting error

It is important to define the point at which equipment stability becomes a significant factor. In general, the total random error will be dependent upon counting statistics, generator and X-ray tube stability and other equipment errors. From the rule for adding variance[5]

$$\varepsilon_{\text{Total}} = \sqrt{(\varepsilon^2)_{\substack{\text{counting} \\ \text{statistics}}} + (\varepsilon^2)_{\text{generator}} + (\varepsilon^2)_{\substack{\text{other equipment} \\ \text{errors}}}} \tag{5.9}$$

By means of, for example, sequential ratio or multichannel spectrometers, the third variable can be reduced to insignificant proportions leaving the counting error and the error due to generator and X-ray tube stability. Where no extra means of overcoming drift is employed, i.e. in high stabiltiy generators, and where the stability of the X-ray tube matches the short term drift of the generator, the point at which the error due to kV and mA fluctuation becomes significant can be estimated. In high stability generators it is common practice to independently stabilize both kV and mA to the same value and the combined instability is usually of the order of three times this value. Thus if both kV and mA are stable to 0.1 % the worst short term stability would be 0.3 %. By substituting this value in Equation (5.7) the number of counts corresponding to this value will be seen to be about 3×10^5. Hence in the case of a generator with 0.1 % stability of kV and mA it is a waste of time to collect more than 3×10^5 counts in a single measurement since in excess of this number, short term drift in the generator becomes the limiting factor.

A similar argument can be applied to any generator X-ray tube combination where the short-term stability is known. It should be appreciated, however, that this drift is short time, i.e. minutes rather than hours, and to make the best advantage of high short-term generator stability ratio measurements must be used. In addition, the X-ray tube output itself should be at least as stable as that of the generator.

Where a precision is required which is numerically better than that allowable by a certain piece of equipment more than one measurement will be required. It can be shown from the rule of variance that if n replicate determinations are made the precision is improved by a factor of $\sqrt{n}$. Thus if 10 measurements of 10^6 counts were made the $\sigma\%$ would be equal to $\frac{100}{\sqrt{10^6}} \times \frac{1}{\sqrt{10}} \simeq 0.03\%$.

5.6 Counting error in the net intensity[8]

In X-ray spectrometry the background is always finite and one of the more difficult problems is to decide whether or not it can be ignored. Fig. 5.5 illustrates a correlation between the true counting error obtained at a peak counting rate of 100 c/s in an analysis time of 120 seconds, against peak to background count ratio. If the background were ignored and the whole of the 120 seconds used to collect counts on the peak, the counting error associated with the 12000 counts taken would be 0.92 %. It will be seen that this value is only approached for very high peak to background ratios, say in excess of 100/1. Fortunately, nearly all of the procedures in general use compare two count rates i.e. from a standard and from an unknown and in these cases the effect of background can usually be ignored, provided the peak to background ratio is in excess of 10/1.

Where background has to be taken into account a rather complex treatment of the peak and background counting rates has to be used in order to ascertain the standard deviation of the net counting intensity.

The standard deviation of the net intensity is derived from the standard deviation of the total intensity $\sigma_p = \sqrt{\dfrac{R_p}{T_p}}$ and the standard deviation of the background measurement $\sigma_b = \sqrt{\dfrac{R_b}{T_b}}$. The standard deviation in the net intensity σ_d (d from difference between R_p and R_b) is $\sigma_d = \sqrt{\sigma_p{}^2 + \sigma_b{}^2}$

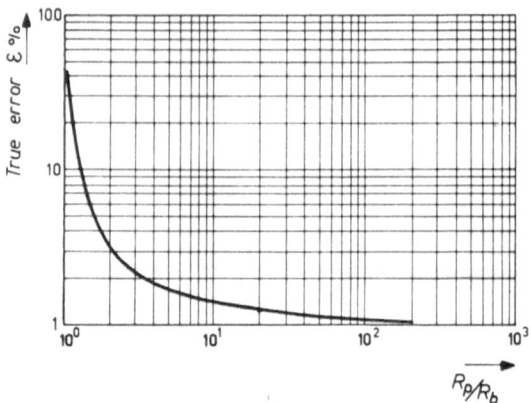

Fig. 5.5 Dependence of the true counting error on the peak to background ratio. If an analysis time of 120 s were used exclusively to record counts from a peak counting rate equivalent to 100 c/s a counting error of 0.92% would result. If, however, allowance were made for the background counting rate to be recorded also within the 120 s analysis time (see Equations (5.14) and (5.16) a curve similar to the one drawn above would result from plotting different peak to background counts ratios. This curve can thus be used as an indication of the extra error which is introduced by ignoring the background.

$$\text{or } \sigma_d = \left| \overline{\frac{R_p}{T_p} + \frac{R_b}{T_b}} \right. \tag{5.10}$$

The relative standard deviation ε in $(R_p - R_b)$ becomes

$$\varepsilon = \frac{\left| \overline{\dfrac{R_p}{T_p} + \dfrac{R_b}{T_b}} \right.}{R_p - R_b} \tag{5.11}$$

This general formula can be adapted for the three most widely used counting methods: fixed time, fixed time optimal and fixed count.

1. Fixed Time
$$T_p = T_b = \frac{T}{2} \text{ with}$$

$$T_p + T_b = T \text{ (total counting time) } \sigma_{F.T.} = \left| \overline{\frac{2}{T}} \right. \cdot \sqrt{R_p + R_b} \tag{5.12}$$

and for the relative standard deviation

$$\varepsilon\% = \frac{100 \sqrt{2}}{\sqrt{T}} \cdot \frac{\sqrt{R_p + R_b}}{R_p - R_b} \tag{5.13}$$

2. Optimal fixed time (cf Section 5.7 and Appendix 4)

$$T_p + T_b = T; \quad \frac{T_p}{T_b} = \left| \overline{\frac{R_p}{R_b}} \right. \tag{5.14}$$

$$\text{then } \sigma_{F.T.O.} = \frac{1}{\sqrt{T}} \cdot (\sqrt{\overline{R_p}} + \sqrt{\overline{R_b}}) \tag{5.15}$$

$$\text{and } \varepsilon\% = \frac{100}{\sqrt{T}} \cdot \frac{\sqrt{\overline{R_p}} + \sqrt{\overline{R_b}}}{R_p - R_b}$$

$$\text{or } \varepsilon\% = \frac{100}{\sqrt{T}} \cdot \frac{1}{\sqrt{R_p} - \sqrt{R_b}} \tag{5.16}$$

3. Fixed count $R_p T_p = R_b T_b = N$

$$\sigma_{F.C.} = \frac{1}{\sqrt{T}} \sqrt{\overline{R_p + R_b}} \cdot \left| \frac{\overline{R_p}}{\overline{R_b}} + \frac{\overline{R_b}}{\overline{R_p}} \right. \tag{5.17}$$

The derivation of the equations for fixed time optimal and fixed count are given in Appendix 4.

The magnitude of the relative errors involved in use of these methods will depend upon the peak to background ratio of the analysis line in question. The following example compares errors obtained applying the three methods where $R_p = 10,000$, $R_b = 100$ and $T = 100$ s.

For $\sigma_{F.T.O.}$

$$\sigma_{F.T.O.} = \frac{1}{\sqrt{T}} (\sqrt{R_p} + \sqrt{R_b})$$

$$\sigma_{F.T.O.} = \frac{1}{\sqrt{100}} (\sqrt{10,000} + \sqrt{100})$$

$$\therefore \sigma_{F.T.O.} = 11$$

For $\sigma_{F.T.}$

$$\sigma_{F.T.} = \left| \frac{\overline{2}}{T} \cdot \sqrt{R_p + R_b} \right.$$

$$\sigma_{F.T.} = \left| \frac{\overline{2}}{100} \cdot \sqrt{10,000 + 100} \right.$$

$$\therefore \sigma_{F.T.} = 14.2$$

For $\sigma_{F.C.}$

$$\sigma_{F.C.} = \frac{1}{\sqrt{T}} \sqrt{R_p + R_b} \cdot \left| \frac{\overline{R_b}}{R_p} + \frac{\overline{R_p}}{R_b} \right.$$

$$\sigma_{F.C.} = \frac{1}{\sqrt{100}} \sqrt{10,100} \cdot \sqrt{100.01}$$

$$\sigma_{F.C.} = 100$$

Thus $\sigma_{F.T.O.} < \sigma_{F.T.} < \sigma_{F.C.}$.

In this example $\sigma_{F.C.} > \sigma_{F.T.} > \sigma_{F.T.O.}$ and indeed this rule generally holds.

This can be proven very simply as follows:

Since count rates are positive, and always $R_p > R_b$, $R_p - R_b > 0$, or

$$(R_p - R_b)^2 > 0, \text{ or } \frac{R_p^2 + R_b^2}{R_p R_b} > 2 \text{ or } \frac{R_p}{R_b} + \frac{R_b}{R_p} > 2, \text{ or}$$

$$\sqrt{\frac{R_p}{R_b} + \frac{R_b}{R_p}} > \sqrt{2} \text{ or}$$

$$\sqrt{\frac{R_p}{R_b} + \frac{R_b}{R_p}} \cdot \sqrt{\frac{1}{T}} \cdot \sqrt{r_p + r_b} > \sqrt{2} \cdot \sqrt{\frac{1}{T}} \cdot \sqrt{r_p + r_b}.$$

Hence $\sigma_{F.C.} > \sigma_{F.T.}$.

For all values of R_p and R_b except for $R_p = R_b$

$$(R_p - R_b)^2 > 0$$

$$\therefore R_p^2 - 2R_p R_b + R_b^2 > 0$$

$$R_p^2 + 2R_p R_b + R_b^2 > 4R_p R_b$$

$$\therefore (R_p + R_b)^2 > (2\sqrt{R_p R_b})^2$$

Or $R_p + R_b > 2\sqrt{R_p R_b}$

and $2(R_p + R_b) > R_p + 2\sqrt{R_p R_b} + R_b$

hence $\sqrt{2} \cdot \sqrt{R_p + R_b} > \sqrt{R_p} + \sqrt{R_b}$

or $\sigma_{F.T.} > \sigma_{F.T.O.}$

This shows that for all values of R_p and R_b (positive) and provided $R_p \neq R_b$ then $\sigma_{F.C.} > \sigma_{F.T.} > \sigma_{F.T.O.}$.

Similarly an optimum time divison has to be considered when the ratio of two counting rates R_1 and R_2 should be determined. The optimum time divison in this case is found to be:

$$\frac{T_1}{T_2} = \sqrt{\frac{R_1}{R_2}}$$

It can be easily demonstrated that the methods of Fixed Count and Fixed Time yield exactly the same precision in the same measuring time when measuring ratios.

5.7 Selection of optimum counting times

Equation (5.16) relates the total time required to obtain a certain value of ε, with the peak and background counting rates. When applying this formula it is necessary to calculate the optimum split of the total time T into the time spent counting the peak response T_p and that on the background T_b. For example, if $R_p = 10,000$ c/s, $R_b = 100$ c/s and an ε of 0.1 % is required, by substitution in (5.16).

$$\sqrt{T} = \frac{100}{0.1} \cdot \frac{1}{100 - 10}$$

This results in $T = 121$ s.

After substitution of the above numerical values of T, R_p and R_b in the relations $T_p + T_b = T$

and $\dfrac{T_p}{T_b} = \sqrt{\dfrac{R_p}{R_b}}$

we obtain $T_p + T_b = 121 \qquad \dfrac{T_p}{T_b} = \sqrt{100}$

and hence $T_p = 110$ s

$\qquad\qquad T_b = 11$ s

In practice it may not be possible to count for precisely 110 and 11 seconds respectively and in general one would select the closest times provided on the scalers of one's own piece of equipment. For example, it might be feasible to measure the peak for 100 s. or 2 minutes (120 s.) and the background for 10 seconds or 0.2 minutes (12 s.).

The error then becomes:

for $T_p = 100$ s and $T_b = 10$ s

$$\varepsilon = \frac{100}{\sqrt{110}} \cdot \frac{1}{\sqrt{10,000} - \sqrt{100}} = 0.106\%$$

if $T_p = 120$ s and $T_b = 12$ s

$$\varepsilon = \frac{100}{\sqrt{132}} \cdot \frac{1}{\sqrt{10,000} - \sqrt{100}} = 0.097\%$$

5.8 Selection of best conditions for analysis

Equation (5.16) can also be used to compare count rates from different equipments or from the same equipment using different operating conditions. It will be seen that for a correctly split fixed analysis time, in order that ε should be as small as possible $(\sqrt{R_p} - \sqrt{R_b})$ should be at a maximum. The value of $(\sqrt{R_p} - \sqrt{R_b})$ is, therefore, often taken as the figure of merit. Application of this figure of merit is far more useful than the sometimes misleading choice of best peak to background ratio as a measure of the optimum operating conditions. The following example compares results obtained, by three different equipments, A, B and C.

Study of the data indicates that equipment A gives the best figure of merit even though equipment C gives the best peak to background ratio.

Equipment	A	B	C
R_p	1600	196	484
R_b	16	4	4
$\sqrt{R_p} - \sqrt{R_b}$	36	12	20
$\dfrac{1}{\sqrt{R_p} - \sqrt{R_b}}$	0.028	0.082	0.050
ε(for $T = 100$ s)	0.28%	0.82%	0.50%
$\dfrac{R_p}{R_b}$	100	49	121

5.9 Selection of best conditions for low concentrations

It can be further shown, that an optimum is also reached when $\dfrac{R_p - R_b}{\sqrt{R_b}}$ is at a maximum, provided that $\sqrt{R_p \cdot R_b} \simeq R_p$.

Since R_p approaches R_b at or near the detection limit, this expression can be used as a quality function for the analysis of very low concentrations. A similar expression is employed in assessing various counting systems to determine low levels of radioactivity.

5.10 Errors in using the ratio method

In many cases use is made of a ratio method in which the time needed to accumulate N counts on a standard S is used as a measuring time for the unknown sample x. The measured value $N_x = R_x T_s$ follows a Gaussian distribution analogous to R_x and $T_s = \dfrac{N_s}{R_s}$. The error ε_R in N_x is:

$$\varepsilon_R = 100 \sqrt{\frac{1}{N_s} + \frac{1}{N_x}} \qquad (5.18)$$

or after substitution of N_s and N_x

$$\varepsilon_R = \frac{100}{\sqrt{T_s}} \sqrt{\frac{1}{R_s} + \frac{1}{R_x}}$$

$$\varepsilon_R = \frac{100}{\sqrt{N_s}} \sqrt{1 + \frac{R_s}{R_x}} \tag{5.19}$$

Provided that R_s is of the same order as R_x

$$\varepsilon_R = \frac{100}{\sqrt{N_s}} \times \sqrt{2}$$

which will be seen to be $\sqrt{2}$ worse than the normal counting error associated with $\sqrt{N}$. (see Equation (5.7))

This would also be predicted from the variance rule since from equation (5.9)

$$\sigma_T = \sqrt{(\sigma^2)_s + (\sigma^2)_x}$$

Provided that R_s is of the same order as R_x, $(\sigma^2)_R \simeq (\sigma^2)_x$

$$\therefore \sigma_T = \sqrt{2(\sigma^2)_s} = \sqrt{2} \cdot \sigma_s = \frac{\sqrt{2} \times 100}{\sqrt{N_s}}$$

Although the counting error is larger in ratio measurements than in absolute measurements the total instrument error can be reduced since the effect of long term drift is eliminated and short term drift only is the limiting factor. Although the magnitude of long term drift is difficult to predict since it depends to a large extent on equipment design and the stability of the mains supply, the authors have found that in general it is of the order of 2-5 times the short term drift.

5.11 Selection of ratio or absolute counting method

The choice between a ratio or an absolute counting method must be made by establishing the total error involved in each method. From Equation (5.9)

$$\varepsilon_{\text{Total}} = \sqrt{\varepsilon^2 \text{ counting} + \varepsilon^2 \text{ equipment}}$$

The errors will be calculated for a typical case in which it is assumed that a mixed programme of absolute and ratio methods is being used in an automatic sequential spectrometer utilizing a standard and at least one unknown sample position. This implies that the standard is measured anyway and no time is saved using absolute measurements. It will be assumed that the equipment error is 0.2 % and that the long term drift is 2.5 times the short term drift.

Case 1 $R_{ps} = 100,000$ cps $R_{px} = 81,000$ cps $T_p = 10$ s

Ratio measurement $\varepsilon_{\text{equip}} = 0.2\%$

$$\varepsilon_x \quad = \frac{100}{\sqrt{1,000,000}} \cdot \sqrt{1 + \frac{100,000}{81,000}}$$

or $\qquad \varepsilon_x \quad = 0.1 \cdot \sqrt{1 + 1.24}$

$$\varepsilon_T \quad = \sqrt{0.04 + 0.0224}$$

$$\varepsilon_T \quad = 0.25\%$$

Absolute measurement $\varepsilon_{\text{equip}} = 0.5\%$

$$\varepsilon_x \quad = \frac{100}{\sqrt{810,000}} = 0.111$$

$$\varepsilon_T \quad = \sqrt{0.25 + 0.012}$$

$$\varepsilon_T \quad = 0.51\%$$

In this case ratio measurement is preferable

Case 2 $\quad R_{ps} = 1000 \quad R_{px} = 810 \quad T_p = 10$ s

Ratio measurement $\qquad \varepsilon_{\text{equip}} = 0.2\%$

$$\varepsilon_x \quad = \frac{100}{\sqrt{10,000}} \cdot \sqrt{1 + 1.24}$$

$$\varepsilon_T \quad = \sqrt{0.04 + 2.24}$$

$$\varepsilon_T \quad = 1.51\%$$

Absolute measurement $\varepsilon_{\text{equip}} = 0.5\%$

$$\varepsilon_x \quad = \frac{100}{\sqrt{8,100}} = 1.11$$

$$\varepsilon_T \quad = \sqrt{0.25 + 1.23}$$

$$\varepsilon_T \quad = 1.22\%$$

In this case absolute measurement is preferable.

5.12 Counting error versus stability

Knowing the properties of counting statistics, it is possible to check the operation of the equipment and observe whether the actual spread in the results agrees with the allowed spread.

From reproducibility measurements N_i the standard deviation $S =$

$$= \frac{\sqrt{\Sigma(N_i - \bar{N})^2}}{k - 1}$$ can be calculated and compared with the allowed

counting error. Normally σ_T must be equal to S if a sufficient number of results is available. The same applies to the relative standard deviation: S/N and ε_T.

This procedure can be used to determine the $\varepsilon_{\text{equip}}$ in the following way: standardise the measuring conditions, choose the lowest possible ε_{st} and accumulate a sufficient number of measurements. If a complete reset is made for each determination the total equipment error will be found; if only some of the measuring parameters are varied the errors associated with the corresponding components can be established.

5.13 Counting error as a function of total number of counts

It should be noted that in many cases the results are not given in count rates R_1 and R_2 but as total number of counts accumulated in a time T.
If the same measuring time is used for peak and background the erorr is

$$\sigma_N = \sqrt{N_p + N_b} \tag{5.20}$$

$$\varepsilon_N = \frac{\sqrt{N_p + N_b}}{N_p - N_b} \tag{5.21}$$

for ratio measurements the error becomes

$$\varepsilon_{r_N} = \frac{1}{\sqrt{N_{st}}} \cdot \sqrt{1 + \frac{N_{st}}{N_x}} \tag{5.22}$$

When using a recorder, pulses are accumulated in an integrating circuit during a time $2(TC)$. The relative standard deviation is then $\dfrac{100}{\sqrt{2R(TC)}} \cdot \%$ in which (TC) is the chosen time constant.

REFERENCES

1 1964 Book of A.S.T.M. Standards, Part 30 pp. 503-511. Tentative recommended practice for the use of the terms precision and accuracy as applied to measurement of a property of a material. A.S.T.M. designation E 177 - 61 T.
2 A Chemists introduction to statistics, Theory of Error and Design of Experiment, 1961, Pantony, Royal Institute of Chemistry Lecture Series No. 2.
3 Philips automatic X-ray spectrometer (P.A.X.) user group, Report of sub committee on the problems of defining the accuracy of X-ray fluorescence analysis, 1965, (M.E.L. Equipment Co., Crawley).
4 PARRISH, Philips Technical Review, 1956, 17, 206.
5 EZEKIAL and FOX, Methods of Correlation and Regression Analysis, Wiley, London, 1959.
6 WILSON and PARRY-JONES, Chemical Analysis, Strouts, University Press, Oxford, 1962.
7 KENDALL and BUCKLAND, A Dictionary of Statistical Terms, Oliver and Boyd, London, 1957.
8 MACK, M. and SPIELBERG, N., 1958, Spectrochim. Acta, 12, 169-178.

CHAPTER 6

MATRIX EFFECTS

6.1 Errors in X-ray analysis

The basis of quantitative X-ray fluorescence spectrometry is to follow the identification of a certain element in a mixture of elements (the matrix) with a measurement of the intensity of one of its characteristic lines, then to use this intensity to estimate the concentration of that element. By use of a range of standard materials a calibration curve can be constructed in which the peak response of a suitable characteristic line is correlated with the concentration of the element. Fig. 6.1 illustrates a typical case where the peak counting rates (R_b) from a range of elements (1-5) are plotted against the concentration of a certain element i. By fitting the calibration curve parameters into the equation for a straight line

$$y \quad = mx + c$$
$$(R_p)i = m_i(\%i) + (R_b)i$$
$$\%i \quad = \frac{(R_p)i - (R_b)i}{m_i} \tag{6.1}$$

it will be seen that the slope of the curve "m" is equal to counts per second per percent and this can be used as a calibration factor for the element in that specific matrix. Once m has been established from standards the net peak minus background response can be divided by m to give the concentration of the element in an unknown but similar matrix. If such a curve were constructed in practice, by an experienced operator using a series of completely homogeneous standards it would be found that, on repeating each measurement a number of times, a certain degree of spread in the count data would occur. This spread is due to certain random errors associated with each reading and would define the precision of the measurement. Random errors arise from a number of sources of which the three most important are (a) Counting Statistics (b) Generator and X-ray Tube Stability, (c) Other Equipment Errors. In addition, it might well be found that even allowing for the expected spread due to random errors, certain data could lay well off the average calibration curve. For example, point 6 in Fig. 6.1 is far too low

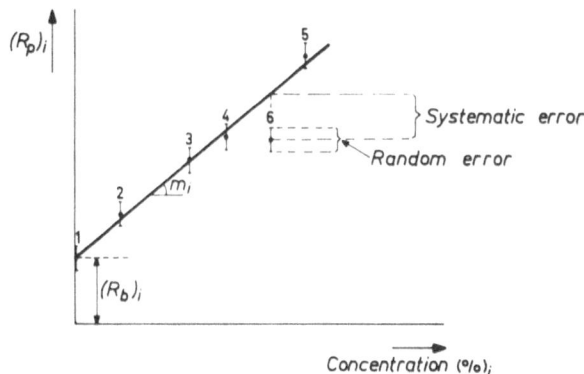

Fig. 6.1 Random and systematic errors in quantitative analysis.

in intensity to fit the average curve. It is apparent that a different type of error is associated with this measurement since a completely different slope factor is associated with it. This error is a systematic error. A typical systematic error might simply be the result of erroneous analytical data on the standard, although, if a standard is truly a standard, by definition its analysis should be accurately known. Assuming all standards to be free from significant analytical errors, the main source of systematic errors in X-ray spectrometry are equipment errors and errors due to the sample itself. Inasmuch as the accuracy of any analysis is dependent upon both associated random and systematic errors, it is important to establish the magnitude of individual errors and to have some control over each of them. From the rule for adding variance[1] the total standard deviation of any measurement will be equal to the square root of the sum of each individual standard deviation squared. Table 6.1 lists the main sources of random and systematic errors and places an order of magnitude on each of them. It is the practice in X-ray fluorescence spectrometry to reduce random equipment errors to such an extent that the error due to counting statistics is the limiting random error within the range of stability of a certain combination of generator and X-ray tube. What we must now concern ourselves with is the control of the systematic errors such that these too can be reduced to such an extent that the counting error is still limiting. Once again systematic equipment errors can be controlled within certain limits by employing suitable design features and provided that the equipment is operated within these design limits, only residual systematic errors due to the sample are important. These residual systematic errors due to the sample are called matrix effects.

TABLE 6.1

Sources of error in X-ray fluorescence spectrometry

RANDOM ERRORS	Counting Statistics (dependant only on time)	
	Generator and X-ray tube stability ($\sim 0.1\%$)	
	Equipment errors ($< 0.05\%$)	
SYSTEMATIC ERRORS	Sample errors	Absorption (100%)
		Enhancement (10%)
		Particle effects (100%)
		Chemical state (5%)
	Equipment errors ($\sim 0.05\%$)	
STANDARD DEVIATION OF MEASUREMENT $(\sigma)^2$	Σ (individual standard deviations)^2	

$$\varepsilon_{total} = \sqrt{\underbrace{\varepsilon^2_{counting} + \varepsilon^2_{generator} + \varepsilon^2_{equipment}}_{random} + \underbrace{\varepsilon^2_{sample} + \varepsilon^2_{equipment}}_{systematic}}$$

Matrix effects fit broadly into two categories — elemental interactions and physical effects. Each of these categories can be broken down further into two sub-sections giving in all four basic types of matrix effects.

These are:

Elemental Interactions   (i) Absorption — Primary and Secondary
                   (ii) Enhancement
Physical Effects   (i) Particle size and Surface effects
                   (ii) Effects due to chemical state

For simplicity, each of the effects will be discussed individually although in practice it is fairly common to find more than one occurring at the same time. In the succeeding chapter on Quantitative Analysis, methods will be discussed for eliminating each of the matrix interactions or at least for reducing them to insignificant proportions.

6.2 Elemental interactions

6.2.1 ABSORPTION

When characteristic radiation is produced from an element in a matrix the number of characteristic photons actually leaving the sample will be significantly less than the number initially produced. This is because most of the excited atoms of the element lie deep within the sample matrix and the characteristic radiation which is produced has to travel through the volume of the matrix in order to leave the sample. The contribution of the outer layers of the sample will be much greater than that of the inner layers and the importance of this effect may be judged from Equation (1.18).

Let us consider the case of a binary mixture AB (Fig. 6.2). A single beam of monochromatic primary radiation of wavelength λ and intensity I_{p_0} enters at point P_1 at an angle ψ_1. The intensity of this beam at point P_2 at depth dp, I_{p2} will be given by

$$I_{p_2} = I_{p_0} \exp - \mu(\lambda)\rho \cdot \frac{dp}{\sin \psi_1} \tag{6.2}$$

where $\mu(\lambda)$ is the absorption coefficient of the matrix and ρ its density. Let us assume that I_{p2} is very small compared with I_{p0}, e.g. that dp is the maximum penetration depth of the primary rays. Characteristic radiation of elements A and B will be excited along $P_1 - P_2$ and will radiate in all directions. The route to the analysing crystal is defined by the primary collimator which makes an angle ψ_2 with the sample surface. The maximum path length x_a of the radiation from element A will be defined by the absorption coefficient μ_A of the matrix for this wavelength.

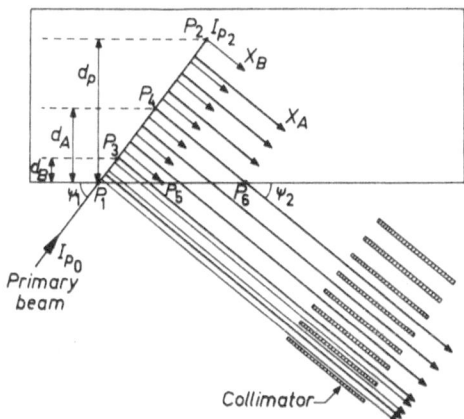

Fig. 6.2 Effect of path length on the volume of sample contibuting to the measured intensity.
Area effectively contributing to intensity of element $A = \triangle P_1 P_4 P_6$
Area effectively contributing to intensity of element $B = \triangle P_1 P_3 P_5$

In general, μ_A will be larger than $\mu(\lambda)$, exceptions being in pure elements, or elements of major concentration where the wavelengths lie close to the absorption edge. In addition to this ψ_2 is generally smaller than ψ_1. Thus, the maximum height d_A from which radiation A can still reach the specimen surface, is significantly smaller than d_p. A similar argument can be applied for element B. Assuming $\mu_B > \mu_A$, the total intensity for A will depend upon the area defined by the triangle $P_1 P_4 P_6$ and for B upon the triangle $P_1 P_3 P_5$. In practice, a broad beam of polychromatic radiation is employed to excite characteristic radiation and the result will be that a certain volume of sample contributes to the effective intensity of each characteristic wavelength. This volume is, in each case, defined by the mass absorption coefficient of the matrix for that particular wavelength. The measured intensity is, of course, a function of this active volume. These active volumes will only be the same for all wavelengths if the absorption coefficients are substantially the same, e.g. if $\mu_A \simeq \mu_B$.

The relative influence of the primary and secondary absorption can be seen in a simplified form of Equation (1.23), i.e.

$$I_j \propto \int_{\lambda \, min}^{\cdot \lambda \, edge} I(\lambda) \cdot \mu_j(\lambda) \cdot \frac{C_j}{\Sigma_i C_i \mu_i(\lambda) + A \Sigma_i C_i \mu_i(\lambda_j)} \cdot d\lambda \qquad (6.3)$$

The secondary absorption term $\Sigma_i C_i \mu_i(\lambda_i)$ can usually be measured or calculated with comparative ease, since discrete wavelengths and absorption coefficients are involved. Primary absorption, however, involves a poly-chromatic beam of radiation and effective wavelengths are sample dependent as discussed in Section 1.3.4. For complex matrices changes in primary absorption are generally less important than changes in secondary absorp-tion as a whole range of wavelengths is involved. The relative influence of secondary absorption can be increased by increasing A, i.e. by working at a low take-off angle. In most spectrometers A has a value of between 1.5 and 2 and the effect of primary radiation can often be ignored. However, this is by no means always the case and it is probably fortunate that the majority of empirical corrections for secondary absorption also correct for primary absorption.

Inasmuch as characteristic radiation is always produced to a finite depth below the sample surface, absorption effects will always be present. The magnitude of these effects will simply be dependent upon the differences in the absorption coefficients of the matrix.

For monochromatic primary radiation Equation (6.3) can be further simplified to:

$$I_j = I_0 \cdot Q_j \cdot \frac{C_j \mu_j}{\mu''} \tag{6.4}$$

where Q_j is a constant and μ'' is an absorption coefficient term containing factors for both primary and secondary absorption [cf 2].

The relative intensities I_1 and I_2 of two different concentrations of the same element measured under identical conditions can be expressed as

$$\frac{I_1}{I_2} = \frac{C_1}{C_2} \cdot \frac{\mu''_2}{\mu''_1} \tag{6.5}$$

Combining this with (6.1)

$$\frac{m_1}{m_2} = \frac{\mu''_2}{\mu''_1} \tag{6.6}$$

where m_1 and m_2 are simply slope factors of calibration curves.

A similar expression will hold for polychromatic radiation if the influences of changes in primary absorption and enhancement effects can be ignored. In this case a "matrix μ" may be defined as

$$\text{matrix } \mu = \Sigma_i (C_i \cdot \mu_i) \tag{6.7}$$

where μ_i is the absorption coefficient of element i in the sample for the secondary radiation. In this case equation (6.6) becomes

$$\frac{m_1}{m_2} = \frac{\mu_2}{\mu_1} \tag{6.8}$$

where μ_2/μ_1 is simply the ratio of the mass absorption coefficient values of the matrix for the secondary radiation only.

Equation (6.5) shows that as the relative ratio of C_1/C_2 increases I_1/I_2 will also increase but at a different rate, now determined by $\dfrac{\mu''_2}{\mu''_1}$. This may in turn lead to positive or negative deviations from a linear relationship. Under very special circumstances, where μ''_2 increases at a relatively greater rate than C_2, even a negative slope may be observed. [3]

Fig. 6.3 illustrates some typical examples of the effect of absorption. Curve AB represents the ideal case where the mass absorption coefficient of the matrix remains relatively constant over the whole range of the mixture. In general a linear relationship will hold provided that the difference in the mass obsorption coefficients over the analysis range does not vary by more than about 5 %. In excess of this value significant deviations do occur giving curves of the type shown by AC. In this instance the element being analysed in the mixture has a higher mass absorption coefficient for its own radiation

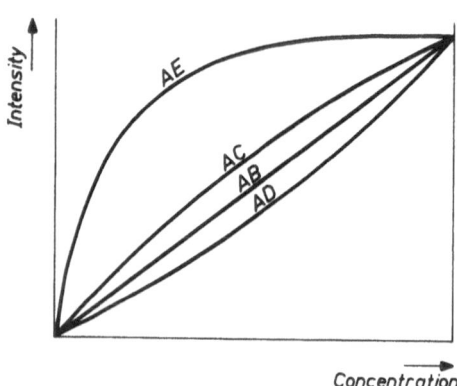

Fig. 6.3 Effect of absorption on the calibration curve.

Curve	Atomic number of element	Average atomic number of matrix	Example	Range of μ
AB	Medium	Medium	Cu(Kα) in CuCO₃/Ca(OH)₂	123—133
AC	Heavy	Medium	Pb (Lα) in Sn/Pb	120—138
AD	Light	Medium	Mg (Kα) in Al/5%Cu	735—435
	Medium	Heavy	Sn (Kα) in Sn/Pb	51— 11
AE	Heavy	Light	T.E.L. (PbLα) in gasoline	2—138

or for the active primary radiation than has the rest of the matrix. The example cited is that of lead/tin mixtures, where the mass absorption coefficient of lead for its own Lα radiation is similar to that for tin Lα, but where the active primary radiation is absorbed far more by lead than by tin. At low concentrations of lead, the tin is the limiting factor in the overall mass absorption coefficient of the matrix, but as more and more lead is added, μ increases causing a decrease of m (cf Equations 6.5 and 6.8) and a gradual flattening-out of the calibration curve.

This effect becomes very marked indeed when the mass absorption coefficient of the analysed element for its own radiation by far and away exceeds that of the matrix (cf Section 1.3.1). This state of affairs is found for example in the case of a very heavy element in a very light matrix. The occurrence of light/heavy element mixtures is surprisingly common in elemental analysis and the many systems fitting into this category include the determination of metals in lubricating oils, petrols, plastics, high alumina or silica minerals and so on. The example quoted (Fig. 6.3 curve AE) is that of tetraethyl lead (T.E.L.) in petrol where the range in mass absorption coefficients is from 2 to 138. Substitution of typical figures in equation (6.7) demonstrates that the matrix mass absorption coefficient increases sharply at first and then less rapidly. For example, the addition of 1 % of T.E.L. to petrol increases the mass absorption coefficient by a factor of two, whereas the addition of 1 % of T.E.L. to a 50/50 mixture of T.E.L. and petrol (in crease in concentration of T.E.L. by 2 % relative) only increases the mass absorption coefficient by a factor of about 1.02. The effect on the m value is considerable — so much so that the calibration curve flattens out completely after addition of about 3 % of T.E.L.

Curve AD is less common and occurs when the mass absorption coefficient of the element for its own radiation is less than that of the remainder of the matrix. Typical examples are found in the determination of magnesium in aluminium/copper alloys and tin in tin/lead alloys.

Similar treatment can be applied to the most complex matrices where absorption effects can be frequently predicted with relative ease from study of wavelength and mass absorption data. [4-7] In general the largest absorption effects are shown by elements fairly close together in the periodic classification owing to the relative proximity of their characteristic lines and absorption edges. Matrix interferences in high alloy steel provide a particularly useful example to illustrate the point, since the majority of elements making up the matrix are of similar atomic number and absorption effects are particularly prevalent. Fig. 6.4 illustrates the problem with specific reference to chrome, manganese, iron and nickel and shows the absorption edges and $K\alpha$ lines which fall between 1.4 and 2.4 Å. To avoid excessive over-crowding

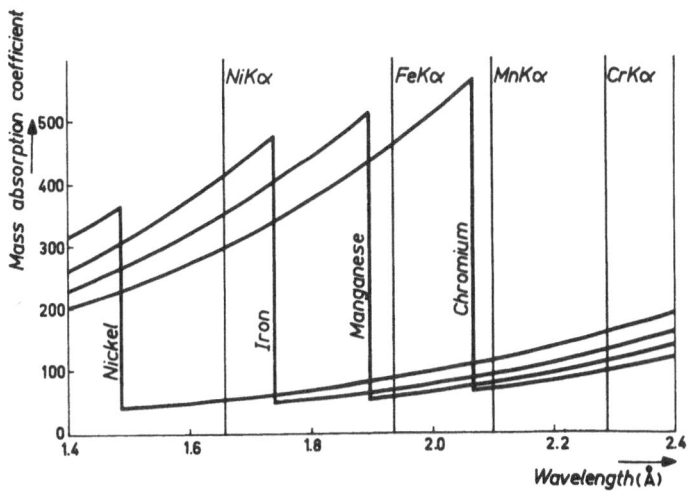

Fig. 6.4 Origin of absorption and enhancement effects in high alloy steels.

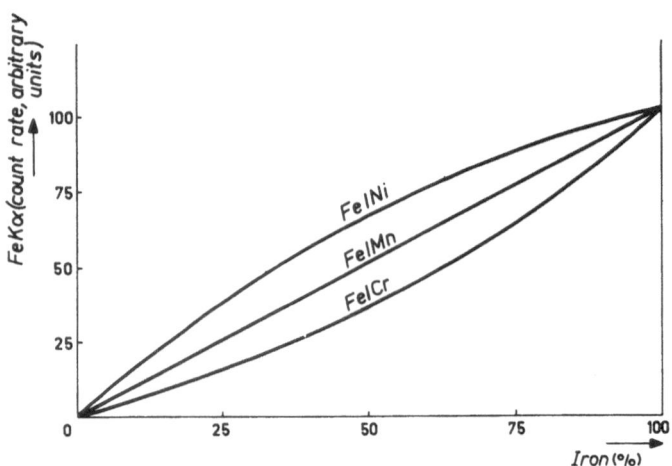

Fig. 6.5 Effect of absorption and enhancement in high alloy steels.

of the diagram the $K\beta$ lines have been left out. It will be seen that in the case of iron and manganese the absorption coefficients are relatively similar, since the iron $K\alpha$ line lies to the long wavelength side of both its own absorption edge and that of manganese. It would be predicted from this that a binary mixture of iron and manganese would show an almost completely linear correlation over the range 0-100 % iron. This effect is illustrated in Fig. 6.5. In the case of iron/chromium however, a rather different situation occurs since the iron $K\alpha$ line lies very close to the chromium absorption edge. Here there is a factor of five difference in mass absorption coefficient for iron $K\alpha$ radiation by chromium and iron. Hence one would expect that as more iron were added to a binary mixture of iron and chromium the absorption coefficient would decrease. The m value would in turn increase giving a calibration curve showing negative deviation as illustrated in Fig. 6.5.

6.2.2 ENHANCEMENT EFFECTS

Where one element is strongly absorbed by another, as for example in the case of iron-chromium already discussed, the measured intensity of the absorbed element will be low by an amount depending on the fraction of photons actually absorbed. In effect a certain amount of energy is held back by the matrix and disposed of by the normal processes of absorption. It will be remembered however, from Chapter 1, that absorption is made up of two components, one of these being due to scattering and the other photo-electric absorption which can give rise to emission of characteristic radiation. Where photo-electric absorption of a characteristic line by a certain matrix element does occur, radiation characteristic of the absorbing matrix element will be produced. This effect is known as enhancement since the number of characteristic photons actually measured is in excess of that predicted due to normal excitation by the primary radiation from the X-ray tube. Thus, in general, where a characteristic line of element A lies just to the high energy side of element B (i.e. where photo-electric absorption is at a maximum) A is said to be strongly absorbed by B and B is in turn enhanced by A. The measured effect in a binary alloy is to produce a positive deviation such as that shown for the iron-nickel curve in Fig. 6.5. Here the nickel Kα line at 1.659 Å lies very close to the iron absorption edge at 1.743 Å, and to its high energy side. The shape of the curve in Fig. 6.5 is determined not only by the enhancement effect, but also by the matrix absorption. The X-rays with the highest absorption probability for iron are those having a wavelength slightly shorter than the iron absorption edge. However, these rays are heavily absorbed by iron, the absorption by nickel for these rays being quite small, (cf equation 1.23) since the absorption edge for iron is situated at 1.74 Å and that for nickel at 1.49 Å. The intensity of iron Kα is, therefore, relatively higher in an iron/nickel alloy than in pure iron. The magnitude of the true enhancement effect depends upon the relative contributions to the total intensity by the primary spectrum and the characteristic sample lines. Its contribution is normally largest either when the element to be enhanced is present in small concentrations, or the element which can enhance is present in major concentrations, or when the primary radiation is not very efficient at exciting the required radiation. Measurements have been made[8,15] to estimate the enhancement contribution and it has been found that even in the most favourable cases, it almost never exceeds 15% of the total measured intensity.

6.3 Physical effects

6.3.1 PARTICLE SIZE AND SURFACE EFFECTS[8-11]

Fig. 6.2 demonstrates that the actual volume of sample which can contri-bute to the measured fluorescent radiation is dependent upon the effective penetration depth of the measured wavelength. This in turn supports a need for a completely homogeneous specimen since if for instance, compositional

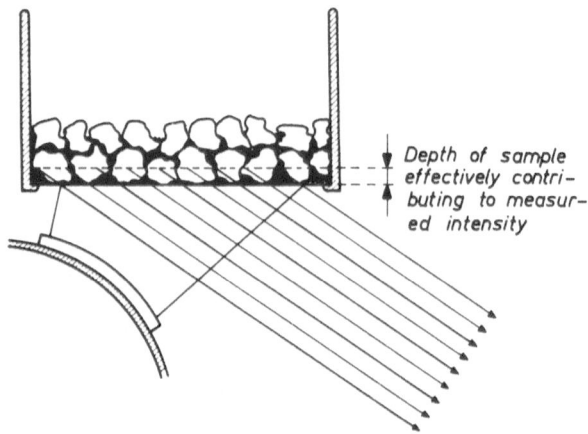

Depth of sample
effectively contri-
buting to measur-
ed intensity

Fig. 6.6 Grain size effects.

variations in depth are present, the measured count data would not be
representative of the whole sample. Such a situation can easily occur in a
badly prepared sample consisting of large and small grains of different
composition. Fig. 6.6 illustrates such a case where the sample consists of
large grains of A along with much smaller grains of B. If the effective depth
of sample contributing to both measured wavelengths is less than the average
particle size of A the effect will be to measure far too much of B and far too
little of A. This effect can only be removed by reducing the grain size of the
whole sample to a value below that of the effective penetration depth. This
in itself can present very great difficulties since the ease with which grain size
can be reduced depends to a very large extent on the physical properties of
the sample in question. In addition the effective penetration depth for the
lower atomic number elements such as magnesium, aluminium and silicon,
in materials such as rock, slags, refractories etc. is only of the order of
5-50 μm and this particle size range can be particularly difficult to achieve.
Just how difficult is demonstrated by the data shown in Table 6.2 which
represents the results of a series of grinding tests[12] made on a syenite sample
containing approximately 15 % Biotite, 15 % Amphibole, 65 % Feldspar and
5 % Quartz. The sample of rock was initially crushed to pass a 20 mesh sieve
and then aliquots of the crushed material subjected to various grinding
techniques. It will be seen that in the cases of both the mechanical agate
mortar and the disc mill significant fractions of the ground sample are present in excess of 50 μm — the majority of these larger particles consisting of
mica. This is fairly typical of the mica like materials since the plate like

TABLE 6.2

Particle Size Distribution

Type of Grinding Technique	Optimum Load	Particles >150 μ	Microscopic Examination	Particles 150 μ—80 μ	Microscopic Examination	Particles 80 μ—50 μ	Microscopic Examination	Particles <50 μ
Mechanical Agate Mortar 2 hrs. Dry Grinding	20 g	2%	Particles up to 300 μ 90% Mica	15%	Rounded Particles 30% Mica	29%		54%
Mechanical Agate Mortar 2 hrs. Wet Grinding	20 g	0.2%	Very large Particles up to 600 μ	1.7%	Rounded Particles 90% Mica	0.4%	60% Mica	97.7%
Tema Disc Mill 12 minutes Dry Grinding	75 g	0.1%	Particles up to 300 μ 95% Mica	1.7%	Rounded Particles 95% Mica	3.3%	40% Mica	94.9%
Glen Creston 270 M & Special Agate Ball Mill. 30 mins. dry grinding	1.6 g	0%	—	<0.01%	a few particles of Mica	<0.01%	few particles of Mica	100%

structure of the crystal inhibits their breakdown. Fig. 6.7 (facing page 65) shows a photomicrograph of a sample ground in a disc mill and clearly demonstrates the wide range in the particle size distribution.

It has also been demonstrated[13) that particle size influences are still present in the case of completely homogeneous specimens where a fall off in intensity occurs as the path length of the measured wavelength approaches the same value as the particle size. This effect is due to mutal shielding of the particles and can be overcome to some extent by sample dilution.[14) A similar mutual shielding effect can be found in the analysis of bulk metal samples where the finite depth of surfacing marks on the analysed face of the specimen may cause count rate fall-off when longer wavelengths are measured. The problems involved in preparing a suitable surface finish in bulk metals is treated fully in the chapter dealing with sample preparation.

Another effect which may influence the measured intensity occurs when the specimen is not completely homogeneous but consists of grains of different chemical composition.[10-11)

For example, suppose that there are two kinds of grain, P and Q, and that element A which is to be measured is only present in the P grains. If the dimensions of these grains are on average, larger than the path lengths of the respective X-rays, then the penetration of the primary X-rays, the exitation of atoms A and the emission of the characteristic X-rays all occur in the P grains only. However, if the grains are smaller than these path lengths, then the absorption occurs in different grains of different composition. This explains why there is a region of grain sizes where the intensity is not very dependent of average grain size. The actual particle size of this region depends on the sample composition and the characteristic wavelength to be measured. A possible way of overcoming this difficulty is to homogenise the specimens, e.g. to work with solutions or glasses (fusion beads). Table 6.3 gives a survey of these effects and indicates the recommended sample preparation.

6.3.2 EFFECTS DUE TO CHEMICAL STATE

Characteristic X-radiation arises following transfer of electrons from outer to inner group orbitals and the wavelength of the characteristic line produced is inversely proportional to the difference in energy between the initial and final states of the transferred electron. For the greater part of the wavelength range normally employed in X-ray spectrometry i.e. 0.1-10 Å the electron is transferred to a K or L group orbital from an L or M group orbital, which in many cases is not the outermost orbital of the atom.

TABLE 6.3

Particle size effects

Composition of grains	Effects to be expected	Recommended practice
All grains same size same hardness same chemical composition	Decreasing particle size leads to increasing amount of material in the effective layer, giving higher intensities	grinding } under constant pressing } conditions
Different composition different hardness different sizes	The softer grains reduce size first, the harder grains break down later during grinding. Intensities for elements in soft grains increase rapidly and decrease later; for elements in harder grains the opposite occurs	grinding to small uniform size, in case of extreme difference in hardness: fusion
Different composition different absorption same or different sizes	Decreasing particle size leads to absorption in different grains depending on path lengths, giving changes in intensity	(Wet) grinding to very small particles ($\approx 5\ \mu m$) to average over many particles; or fusion.
All elements rather homogeniously distributed over the grains		reproducible grinding and pressing
A given element is only present in a certain compound		fusion

Since changes in electron density due to valence and/or co-ordination are associated with the outermost or next to outermost orbitals significant wavelength shifts due to variations in chemical composition are the exception rather than the rule. However, this is not always the case and wavelength shifts can sometimes present matrix problems which are particularly difficult to overcome. As an example, any element below atomic number 18 has unfilled $3p$ orbitals and since the $K\beta$ line arises following a transition from this level it is to be expected that the wavelength of the $K\beta$ line for elements below atomic number 18 be dependent on chemical state. This is indeed true and it is also true to a lesser extent for the $K\alpha_1$, and $K\alpha_2$ lines which arise from $2p$ transitions. Fortunately the number of elements in this range is relatively small, i.e. sodium (Z = 11) to chlorine (Z = 17), and only two or three of these occur regularly in different states of oxidation or co-ordination — these being sulphur, aluminium and to a lesser extent silicon.

Sulphur is an element which occurs in a wide range of oxidation states ranging from 2^- in sulphide ion to 6^+ in sulphate ion. Fig. 6.8 shows the result [15] of a step scan over the sulphur $K\beta$ line measured from a sample of sodium thiosulphate which contains both sulphur 6^+ and sulphur 2^-. Two maxima are seen — the difference in wavelength corresponding to approximately 0.0015 Å or about 2 eV. This value is however significantly lower than the figure of 6.5 eV obtained by Hagstrom et al[16] using the rather more sensitive technique of electron spectroscopy.[17]

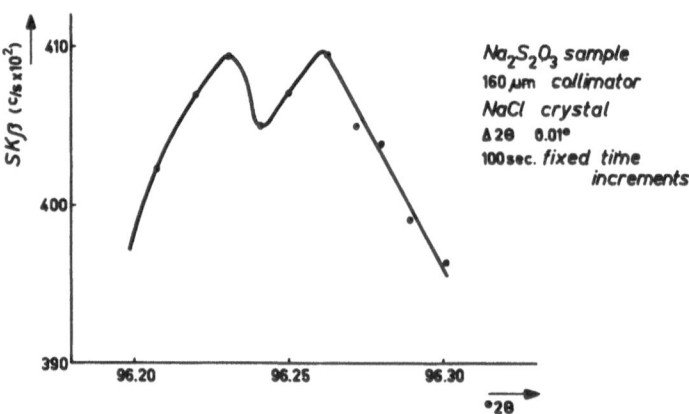

Fig. 6.8 Chart record obtained by step scanning over the SK β line from a sample of sodium thiosulphate. Since both S^{2-} and $S^{6\pm}$ are present two peaks occur corresponding to the two wavelengths of sulphur.

Aluminium occurs regularly in three different states of co-ordination viz. 6, 4, and 0 (e.g. as in aluminium metal) and Fripiat and his co-workers[18] have shown a roughly linear relationship between wavelength and co-ordination number, with a maximum difference of 0.013 Å between Al^{VI} and $Al°$. The same outhors have shown that the maximum wavelength difference exhibited by silicon is 0.009 Å and have further demonstrated the usefulness of wavelength shift to determine co-ordination numbers of silicon and aluminium in samples of slags and zeolites. A combination of X-ray line shift and infra-red spectroscopy has also been employed to study the structure and properties of amorphous silicoaluminas.[19] The $K\alpha_3$ and $K\alpha_4$ lines are also particularly sensitive to electron density in the outer orbitals since these lines arise from dual transitions. Unfortunately these lines are rather too weak to be of much practical use although some data has been published.[20] As would be expected wavelength shifts are particularly prevalent in the ultra-soft X-ray region (20-80 Å) but since this wavelength range is still considered to lie outside the normal working range of the conventional X-ray spectrometer it is not the intention to discuss this region any further. The reader is advised to consult the literature dealing exclusively with this topic.[21-23]

Wavelength shifts are by no means restricted to the low atomic number region since several ranges of elements are typified by their partial filling of penultimate principle orbital groups. For example since the transition elements have partially filled $3d$ orbitals the $K\beta$ wavelengths for this series of elements would be expected to be co-ordination/valence dependent. This is indeed the case and several groups of workers have attempted to make use of this shift for the determination of valence/co-ordination state in the transition metal series.[24] For example, Zemany[25] has used a conventional X-ray spectrometer to measure the wavelength difference between manganese $K\beta$ in $KMnO_4$ and $MnSO_4.H_2O$. His value of 0.001 Å correlates reasonably well with the data cited by Fugge[26] in his summary of the K wavelength shift data for transition metal oxides. The second long period elements with partially filled $4d$ orbitals would similarly be expected to show wavelength shifts in their $L\beta$ lines and Haglund[27] has reported shifts up to 0.006 Å in the most favourable cases. Table 6.4 lists some of the wavelength shift data that have been reported and it is immediately noticeable that the degree of shift decreases sharply with atomic number within a given spectral series. This is, however, to be expected since the maximum wavelength change is dependent upon the largest possible difference in energy of the transferred electron i.e. on its potential energy before transference. For example, in the case of $K\beta$ lines

$$1/\lambda \propto E_K - E_M \tag{6.14}$$

where E_M and E_K represent the initial and final states of the transferred electron. Since for a given element E_K is virtually constant the magnitude of the wavelength is dependent solely upon E_M. Thus aluminium, which can show a maximum M shell electron density increase of $3 \to 8$, can give $K\beta$

TABLE 6.4

Effect of electron configuration on wavelength

Element	Atomic Number	Electronic Configuration	Possible Electron Gain	Maximum Wavelength Shift (Å)
Aluminium	13	K, L, $3s^2 3p^1$	3-8	0.0013[15]
Silicon	14	K, L, $3s^2 3p^2$	4-8	0.0009[15]
Sulphur	16	K, L, $3s^2 3p^4$	6-8	0.0015[19]
Chromium	24	K, L, $3s^2 3p^6 3d^4, 4s^2$	12-18	0.0004[28]
Manganese	25	K, L, $3s^2 3p^6 3d^5, 4s^2$	13-18	0.0003[28]
Cobalt	27	K, L, $3s^2 3p^6 3d^7, 4s^2$	15-18	0.0001[28]
Nickel	28	K, L, $3s^2 3p^6 3d^8, 4s^2$	16-18	0[28]
Zirconium	40	K, L, M, $4s^2 4p^6 4d^2, 5s^2$	10-18	0.005[21]
Niobium	41	K, L, M, $4s^2 4p^6 4d^4, 5s^1$	12-18	0.006[21]
Molybdenum	42	K, L, M, $4s^2 4p^6 4d^5, 5s^1$	13-18	0.005[21]
Rhodium	45	K, L, M, $4s^2 4p^6 4d^8, 5s^1$	16-18	0.001[21]

wavelength shifts of up to 0.013 Å whereas cobalt with a maximum possible electron density increase of $15 \to 18$ shows only an 0.0001 Å change. It is also apparent from the data in Table 6.4 that $L\beta$ shift is much more marked than $K\beta$ shift and this is only to be expected since the energy gap between $E_M - E_K$ is much greater than $E_N - E_L$ and hence the relatively small energy increment involved in the addition of one electron is more significant in the case of the $E_N - E_L$ transition.

The electron distribution in the outer orbitals determines not only the wavelengths of characteristic lines but also their relative intensities. The relative intensity of, for instance a $K\alpha$ plus a $K\beta$ line depends on the probability of an electron jump from the L or M orbital. This probability depends in its turn on the energy difference between the respective orbitals, which can be expressed as a relation in which the energy difference is in an exponential term. It is thus obvious that the intensity ratio should be more sensitive to changes in the outer electron distribution than for example the wavelength shift. It is, however, rather difficult to measure the intensities accurately as the $K\beta$ and $K\alpha_3$ lines are rather weak. It is also necessary to correct the measured intensities for the different absorption in the sample.

Owing to the relatively crude collimation employed in X-ray spectrometry, especially in the case of the longer wavelengths, line shifts only become a problem where the displacement exceeds about 0.001 Å. In practice this means β lines of greater than around 4 Å and α lines in excess of about 7 Å. Conversely where line shift values are required as a measure of co-ordination/valence effects, care must be taken to work under conditions of high resolution i.e. smallest possible "d" spacing crystal and highest possible degree of collimation, and also to carefully temperature control the analysing crystal.[19] Where information is required on the chemical state of elements with unfilled penultimate orbitals recourse is best made to the use of absorption edge fine structure measurements,[25, 28, 29] or to electron spectroscopy.[17]

REFERENCES

1 PANTONY, D. A., 1961, Royal Institute of Chemistry Lecture Series, No. 2.
* 2 BIRKS, X-*Ray Spectrochemical Analysis*, New York, Interscience, 1959.
3 BIRKS, L. S. and HARRIS, D. L., 1962, Analyt. Chem., **34**, 943.
* 4 MITCHELL, *Encyclopedia of Spectroscopy*, Reinhold, New York, 1960, 0. 736.
5 MITCHELL, B. J., 1961, Analyt. Chem., **33**, 917.
6 GLOCKER, R. and SCRIEBER, H., 1928, Ann. Physik., **85**, 1089.
7 LEIBHAFSKY, H. A., 1954, Analyt. Chem., **26**, 26.
8 ADLER, I. and AXELROD, J. M., 1955, Spectrochim. Acta, 7, 91.
* 9 CLAISSE, F., 1957, Norelco Reporter, **3**, 3.
10 CLAISSE, F. and SAMSON, C., 1962, Report S67, Province de Quebec, Service des Laboratoires.
11 BLANQUET, P., *L'analyse par spectrographie et diffraction de rayons* X, (Madrid, 1962), Philips, Eindhoven, 85.
12 POOLE, A. B., 1966, Private communication.
13 GUNN, *Advances in X-ray analysis*, Plenum, New York, 1960, **4**, 382.
14 GLOTOVA, A. N., LOEV, N. F. and GUINICHEVA, T. I., 1964, Ind. Lab., **30**, 863.
15 DE VRIES, J. L., *Proceedings of XII International Spectroscopy Colloquium*, Hilger & Watts, London, 1965.
16 HAGSTROM, S., NORDLING, C. and SIEGBAHN, K., 1964, Z. Physik, **178**, 439.
17 NORDLING, C., HAGSTROM, S. and SIEGBAHN, K., 1964, Z. Physik, **170**, 433.
18 FRIPIAT, J. U., LEONARD, M. and DE KIMPE, C., *Analyse par les rayonnements* X, (Bruxelles, 1964) Philips, Eindhoven.
* 19 LEONARD, A., SUZUKI, SHO, FRIPIAT, J. J. and DE KIMPE, C., 1964, J. Chem. Phys., **68**, 2608.
20 BAUN, W. L., and FISCHER, D. W., 1964, Nature, **204**, 642.
21 BAUN, W. L., FISCHER, D. W., 1965, Analyt. Chem., **37**, 902.
22 FISCHER, D. W. and BAUN, W. L., 1964, A.F. Materials Laboratory Report, No. RDT-TDR-63-4232.
* 23 FISCHER, D. W., and BAUN, W. L., 1965, J. Chem. Phys., **43**, 2075.
* 24 WHITE, E. W., MCKINSTRY, H. A. and BATES, T. F., *Advances in X-ray analysis*, Plenum, New York, 1958, **2**, 239.
25 ZEMANY, P. D., 1960, Analyt. Chem., **32**, 595.
26 FLUGGE, *Encyclopedia of Physics*, Springer, Berlin, 1957, Vol. **30**, p. 156.
27 HAGLUND, P., 1941, Arkiv. Mat. Astrom. Fysik. Ser. A28, No. 8.
28 VAN NORDSTRAND, R. A., 1960, Advances in Catalysis, **12**, 149.
29 LEVY, R. M., 1965, J. Chem. Phys., **43**, 1946.

CHAPTER 7

QUANTITATIVE ANALYSIS

7.1 General

The preceding chapters have discussed the various random and systematic errors which can arise during an analysis either from the equipment or from the sample to be analysed and it is the purpose of this section to discuss the methods which are available for reducing these errors to an acceptable value. The first problem is however, to define an acceptable value. Instrumental techniques are invariably adopted for one reason only, this being the inherent speed of the instrumental method compared to classical wet techniques. Since practically all analytical instruments are nothing more than rapid, versatile and frequently very expensive comparison devices, recourse has nearly always to be made to the use of chemically analysed or synthesised standards. Basically no data obtainable by an instrumental technique can be more accurate than that of the standards with which they were compared, although the random errors of the chemical analysis can be reduced by graphical or mathematical interpolation. A calibration graph is thus inherently more precise than its individual points. It is frequently impracticable to start with basic measurements of weight, volume etc. and one must invariably accept either the chemically or synthesised standard as the ultimate limit to the accuracy obtainable, or settle for very precise trend information which may or may not be of comparable accuracy.

The limits of accuracy obtainable with chemical methods of analysis is inevitably a point of some conjecture and usually the best and most reliable data are those available from multi-laboratory, multi-sample (round robin) analytical tests. Typical of the results of such tests are those taken from a recent programme concerning British Standard Methods for the analysis of iron and steel samples [48].

Data showed an approximately square root relationship between standard deviation s and determined concentration C; in fact $s = K\sqrt{C}$. For the 95% confidence limits and duplicate determinations the factor K was found to lie between 0.008 and 0.054. It is therefore reasonable to assume that the accuracy of a good chemical analysis be of the order of 0.2% for a duplicate

TABLE 7.1

Methods of Quantitative Analysis	Corrects for				
Method	Absorption	Enhacement	Particle Size	Density	Long term Equipment Drift
Standards					
a) External (limited range using similar materials)	√	|	|	|	|
b) Internal (different element)	|	|	—	—	|
c) Internal (same element)	|	√	—	—	|
d) Scattered tube line (coherent)	|	—	—		|
e) Scattered tube line (incoherent)	|	—	—	(|)	|
Dilution					
a) Addition of a large quantity of low absorber	|	|	(|)	|	—
b) Addition of a small quantity of high absorber	|	√	—	—	—
c) Liquid solution	√	|	|	|	—
d) Solid solution	|	|	√	|	—
Pelleting					
a) Low pressure (1-5 tons sq. in.)	—	—	—	|	—
b) High pressure (∼50 tons sq. in.)	—	—	(|)	|	—
Mathematical Correction					
a) Empirical	√	|	—	—	—
b) Semi-Empirical	√	√	—	—	—
c) Correction using data obtained by other measurements	√				—
Thin Film	√	√	—	—	—

determination, thus a reasonable target to aim for using an instrumental technique is a coefficient of variation (i.e. $\sigma \%$ see chapter 5) of around 0.15%. The modern X-ray spectrometer utilizing some form of ratio or reference analytical procedure can achieve this precision fairly comfortably and the task of the X-ray spectroscopist is to device a quantitative method which will reduce matrix effects such that they too are at or below this value.

The range of methods and procedures which have been adopted in X-ray fluorescence spectrometry is enormous, a statement easily verified by study of the 300 or so papers which are published annually dealing with applications of the technique. This chapter deals only with the more important of these and the reader is recommended to study the excellent review articles which are published regularly[1-2] for further information about applications in his or her particular field.

The procedures which are adopted for the development of a method of quantitative analysis can be conveniently broken down into different categories. Table 7.1 shows a typical list and indicates which of the five main causes of systematic error each procedure will eliminate. It will be seen immediately that no one method can be universally applied and one must frequently choose a combination of procedures to suit any particular situation. In order to overcome the effects of long term instrumental drift comparison must always be made with some form of standard, be it in the form of solid block of material kept permanently in one of the analysis positions, or perhaps as a special reference channel built into the equipment. As far as matrix effects are concerned it will be seen that although many methods are available for overcoming absorption and enhancement effects only solution techniques will overcome all four problems at the same time. Only when one is working with limited concentration ranges in a series of chemically similar materials is it possible to normalize the experimental conditions to such an extent, that these disturbing effects can be taken care of by the choice of standards. In certain instances the listed method may go a long way to minimizing a particular interference but may not overcome it completely. Such cases are indicated by placing the tick, indicating successful removal of a particular effect, in parentheses.

7.2 Use of standards

7.2.1 EXTERNAL STANDARDS

One is frequently faced with the problem of analysis for a particular element which is present in a series of samples over a fairly limited concentration range. It will be seen from figure (6.3) that although matrix effects may be such that a wide deviation from linearity occurs over a large concentration range, suitable ranges are available where to all intents and purposes the count rate concentration correlation can be considered to be linear. It is of course also necessary to have sufficient sensitivity (i.e. rate of change of count rate with concentration) and whereas the low concentration ranges of curve AE could be successfully employed in this context, the high concentration ranges are completely unuseable. It is difficult to give a hard and fast rule as to the range over which absorption and enhancement effects can be considered negligible, but in very general terms if the range in mass absorption coefficient for the required wavelength does not exceed 5 % or so and provided that variation in the concentration of possible enhancing elements is also less than 5 %, linear curves should be obtained. Where these circumstances exist calibration curves covering the required range can be constructed by use of external standards which have been chemically analysed. This situation is frequently met in the analysis of bulk metals where minor elements (up to about 5 %), or close concentration ranges of a major element, can be successfully analysed by this means, provided that care is taken to prepare both the sample and standards in the same manner. In the case of powders great care must be taken to ensure that sample and standards have the same particle size distribution since systematic differences are by no means uncommon between routine samples and standards which have been synthetically prepared from commercially available chemicals. Such chemicals are frequently prepared by precipitation and in consequence have a very fine particle size. Particle size and density variations can often be minimized by high pressure pelletizing of the sample, a technique which is discussed in the following chapter dealing with sample preparation. This relatively simple procedure will often give excellent results in quite complex matrices[3] provided that care is taken in the reproducibility of the pelletizing technique.

Solutions can often be analysed with great ease by direct comparison with external standards owing to the simplicity of preparing standard solutions. It may be worth pointing out that a simple check on the absence of enhancement or absorption effects can be made by diluting the highest solution standard in the range by a factor of two with pure solvent, which should in turn drop the net peak response by the same amount.

7.2.2 INTERNAL STANDARD (DIFFERENT ELEMENT)[4]

Since absorption and enhancement effects are caused by interferences arising from the close proximity of characteristic lines and absorption edges one possible way of overcoming them is to add an internal standard having a wavelength which is affected in a similar manner as the analysed wavelength. An example might be the determination of iron in Fe_2O_3/Cr_2O_3 mixtures illustrated in Fig. 7.1. Here the iron $K\alpha$ line at 1.937 Å is strongly absorbed by chromium whose absorption edge is at 2.070 Å.

Cobalt has its characteristic $K\alpha$ line at 1.791 Å and since the absorption of iron and cobalt by chromium is similar a weighed quantity of, for example, Co_2O_3 could be added to the Fe_2O_3/Cr_2O_3 mixture as an internal standard.

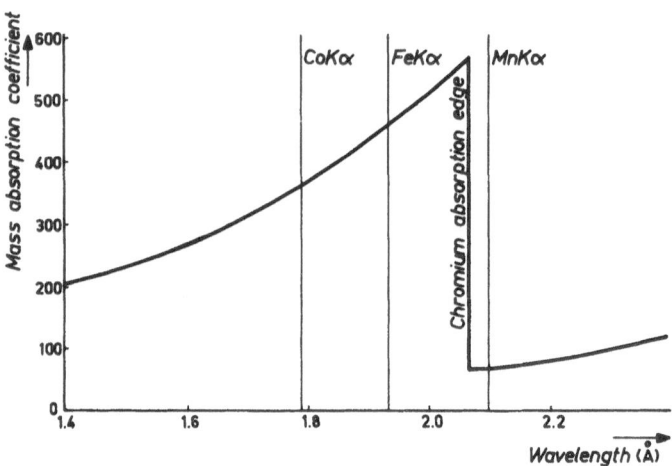

Fig. 7.1 Choice of internal standards. Fe Kα (1.937 Å) is strongly absorbed by chromium. The Mn Kα line (2.103 Å) is no use as an internal standard because it lies to the long wavelength side of the chromium absorption edge (2.070 Å). Co Kα is a far better choice because it lies to the short wavelength side of the chromium absorption edge, hence the absorption coefficients for Co Kα and Fe Kα by chromium are similar.

Since absorption effects for iron and cobalt are similar and as the concentration of cobalt is known

$$\frac{\text{Count rate due to FeK}\alpha}{\text{Count rate due to CoK}\alpha} = K \cdot \frac{\text{Concentration of Fe}}{\text{Concentration of Co}}$$

The calibration factor K could be determined from standard samples of $Fe_2O_3/Cr_2O_3 + Co_2O_3$ and then this factor used in turn to determine iron concentrations in unknown mixtures. It is assumed in this method that particle size and density effects are negligible and particular care must be taken to ensure that the particle size distribution of sample and internal standard are similar. This is why the internal standard method is particularly useful dealing with solutions as they allow a completely homogeneous mixing of sample and standard.

It will be apparent from the study of wavelength tables that the best choice of internal standard is to be made from elements of atomic number similar to that of the analysed element. It is essential, however, to ensure that the wavelength of the internal standard is at the same side of the absorption edge of the absorbing element as the analysed wavelength. For example, cobalt ($Z = 27$) is a good internal standard for iron ($Z = 26$) but manganese

from which C_1 can be deduced. This method, which is sometimes known as spiking, is extremely useful for the determination of single elements in very complex matrices since nothing at all need be known about other elements which are present in the sample. In order to obtain reliable results it is necessary that C_2 be of the same order as C_1, otherwise the experimental spread in the measurement of R_p may influence the slope of the curve, which can lead in turn to erroneous values of C_1. In order to avoid tedious calculations, the actual weight of added substance should be small compared with the weight of total sample. This is because addition of the amount C_2 to the sample slightly lowers the original value of C_1. This is one of the reasons why the method of spiking is limited to rather low concentrations, say below 5%. If the correlation between count rate and concentration is not linear, the actual shape of the calibration line has to be established by repeating the addition, if need be several times. In any event it is best to make a second addition in order to check the linearity of the calibration curve. Again great care must be taken to ensure that the particle size and density of the added element is similar to that of the sample.

7.2.4 USE OF SCATTERED TUBE LINES

The sample will emit not only fluorescence lines characteristic of the elements present but will also scatter the primary X-ray tube spectrum. This scattering may be coherent i.e. when the scattered line has the same wavelength as the primary line (Rayleigh scatter), or incoherent when the scattered line has a slightly longer wavelength (Compton scatter). The relative intensities of the excited and scattered lines are equally dependent upon sample positioning and X-ray tube current. Hence the intensity ratio of a characteristic line to a scattered X-ray tube line is rather insensitive to instrumental errors which arise from these sources. As the scattering power of a matrix is related to its average atomic number and its density, it should be possible to select a scattered tube line, or a part of the scattered continuous spectrum, to act as an internal standard to correct for absorption and density effects. Different relationships have been proposed[10-14] which correlate the scattered intensity either with the matrix absorption of the sample, or directly with the emitted fluorescence intensity. For a given element and a given instrument, the fluorescence intensity I_f is proportional to $1/\mu$, where μ is the matrix mass absorption coefficient (including both primary and secondary absorption) i.e. proportional to Z^{-4}. The intensities of the coherently scattered tube lines are proportional to the square of the scattering power of the matrix, which is in turn dependent upon Z, and inversely proportional to μ, i.e. to Z^4. The scattered intensity I_b will thus be proportional to Z^{-2} and

$$I_b/I_f \propto Z^2 \tag{7.2}$$

This relationship only holds when the fluorescent and scattered lines under consideration have substantially the same wavelengths. This ratio is less dependent upon the matrix composition than is I_f alone. It is necessary to consider that the total scattering power function is determined by the summation of the contributions of the separate atoms in a certain way and the matrix μ by summing the absorption of the atoms in a different way. Since both scattering power and absorption are strongly wavelength dependent, it is possible to find background positions (λ_b) where the ratio I_b/I_f is almost completely insensitive to changes in matrix composition. It is useful to note that I_b, being proportional to Z^{-2}, increases with decreasing matrix atomic number and it follows that for low atomic number matrices, such as aqueous solutions and hydrocarbons, the background is intense and thus it is much easier to measure line intensities, tube lines or scattered continuum, than in the case of specimens composed of higher atomic number elements.

The intensity of the incoherent scatter is also dependent upon the primary wavelength. The compton scatter increases with decreasing atomic number and with decreasing wavelength. Both coherent and incoherent scatter depend upon the electronic structure of the atoms, but in quite a different way. The incoherent scatter[13] and the coherent scatter[10-12] have both been used to correct for variations in mass absorption coefficient and instrumental drift. The intensity ratio of coherently to incoherently scattered characteristic tube lines has been used to determine carbon to hydrogen ratio using both $WL\alpha$[14-15] and $Mo\ K\alpha$[16] primary radiation. As the scattering power of a matrix is very dependent upon its composition, it has also been suggested that the coherent/incoherent scatter ratio could be used for the chemical analysis of light elements in a moderately light matrix[16].

7.3 Dilution techniques

It will be apparent from the general intensity formula and especially from Equation (1.26)

$$C(\lambda \lambda_j) = \frac{\mu_j(\lambda)}{\Sigma_i C_i \mu_i(\lambda) + A \Sigma C_i \mu_i(\lambda_j)} \tag{1.26}$$

that changes in the slope of the calibration curve (Δm) depend upon changes in the total value of the denominator resulting from variations in the composition. These deviations will be most pronounced when the concentration (C_a) changes for an element a whose absorption coefficient $\mu_a(\lambda)$ or $\mu_a(\lambda_j)$ differs very markedly from the other μ values. Variations in the denominator will be insignificant when the contribution of $C_a \cdot \mu_a$ does not alter the total value of the denominator to any great extent. This situation may be brought about by adding a diluent d to the sample, such that $C_d \cdot \mu_d$ is large compared with $C_i \cdot \mu_i$. The following equation then holds:

$$C(\lambda \lambda_j) = \frac{\mu_j(\lambda)}{C_i[\mu_i(\lambda) + A\mu_i(\lambda_j)] + C_a[\mu_a(\lambda) + A\mu_a(\lambda_j)] + C_d[\mu_d(\lambda) + A\mu_d(\lambda_j)]}$$

(7.3)

The last term of the denominator will be much larger than the others either if C_d is very large — this means a high degree of dilution, or if $[\mu_d(\lambda) + A\mu_d(\lambda_j)]$ is large, this in turn entails the adding of a highly absorbing material to a low absorbing matrix. For example, when measuring Si Kα in mixtures of SiO_2 and Fe_2O_3, the spread in matrix mass absorption coefficient is 660-1804, a range of 170%. By diluting 9 : 1 with a low absorber, such as lithium carbonate, this range could be reduced to about 16%. The same reduction could, however, also be achieved by a 2 : 1 dilution with a heavy absorber, such as lanthanum oxide.

Since the degree of enhancement is also proportional to the concentration of the enhancing element, it is to be expected that addition of an inert diluent to the matrix would reduce enhancement effects as well.

The required degree of dilution can be estimated from Equation (7.3) since deviations in (Δm) are directly proportional to deviations in $C(\lambda \lambda_j)$. It can be stated in general terms that linear calibration curves will be obtained provided that the range in matrix mass absorption coefficient is reduced to about 5%.

Dilution methods have been employed with great success over a wide range of applications but probably have been most successful in the analysis of geological samples[17-18)] where elemental concentration ranges can be enormous and where the large number of elements precludes the use of internal standards.

One disadvantage of the dilution method is that complete mixing is absolutely vital. This degree of mixing is best achieved by using the diluent as a solvent for the sample, either as aqueous or acid solutions or by making solid solutions by fusion. In this way it is possible to remove not only the effects of absorption, enhancement and density but also to remove intensity variations attributable to particle size effects. The choice of solvent will of course be determined by the chemical properties of the sample concerned and typical examples are given in the appropriate section on sample handling. Although many methods of matrix stabilisation by dilution have been successfully employed all suffer from the same fundamental disadvantage — this being that the dilution necessarily reduces the number of analysed atoms per irradiated volume with a corresponding loss in intensity. This can be particularly troublesome in the analysis of low concentration of light elements where sensitivities may already only be marginally useful.

7.4 Thin film techniques

Absorption effects within the analysed sample become less as the sum of the path lengths of the primary and secondary radiation decrease. In general terms it can be said that the self absorption of the sample becomes negligible where

$$[\mu_p^x \cdot \text{cosec } \psi_1 + \mu_s^x \cdot \text{cosec } \psi_2]x \ll 1 \qquad (7.4)$$

μ_p^x and μ_s^x are the linear absorption coefficients of the specimen for primary and secondary radiation and ψ_1 and ψ_2 are the incident and take-off angles of the spectrometer. Provided that the relationship given in Equation (7.4) holds the intensity from any element is directly proportional to its concentration, thus for any two weight concentrations C_1 and C_2 of the same element

$$\frac{C_1}{C_2} = K \cdot \frac{I_1}{I_2} \qquad (7.5)$$

where I_1 and I_2 are the corresponding intensities. K is a constant for any one element measured under specified instrumental conditions. This technique was originally devised for the analysis of thin metallic films of iron, chromium and nickel[19] and more recent work[10] has demonstrated that by means of careful calibration procedures very accurate data can be obtained in the thickness range 30-6000 Å. Other workers[21] have demonstrated that by deliberately converting a sample to be analysed into a thin layer all the advantages of this technique can be gained with only marginal loss of sensitivity. A thin film can be prepared quite simply by evaporating a solution of the unknown onto mylar film[22] or a filter paper disc[23] and should an internal standard be required this can be added very simply by spotting a known volume of standard solution on to the disc by means of a micro-pipette. One of the disadvantage of this technique is that since the available intensity is decreased, the determination of minor concentrations is rarely practicable. Also it is very often difficult to prepare thin films of constant thickness and thus accurate quantitative analysis is only possible if all elements are determined and if the variation of intensity with thickness is similar for all these elements.

7.5 Mathematical corrections

Since it is possible to go a long way to relating the characteristic line intensity to the concentration of a given element, in terms of measurable parameters, it is not suprising that since the early days of X-ray fluorescence spectrometry attempts have been made to employ mathematical matrix correction procedures.[24-29] However the parameters relating concentration and line intensity frequently produce rather complex functions (see Section 1.4.3) with the result that alternative semi-empirical correction procedures have also been developed which are usually far easier to apply and frequently produce data which is at least as accurate as that obtained from more sophisticated methods. More recently the growth in the development of computer technology has given new impetus to the application of mathematical correction procedures and there seems little doubt that more and more workers will turn to this approach for the analysis of multielement matrices in preference to the more time consuming standard addition or dilution techniques. Unlike standard addition and dilution techniques, however, mathematical correction methods will not overcome particle size effects and this must be born in mind particularly in the analysis of light elements in powder samples.

The first real attempt to derive working equations from first principles was made as long ago as 1955 by Sherman[30] and although the same author later provided a somewhat simplified version[31] there has been little further attempt to follow this line. With certain exceptions[32] the majority of workers have preferred to employ purely empirical relationship[33] mostly based on multiple regression methods[34-36] or influence factors[37-40].

7.5.1 PRINCIPLE OF THE INFLUENCE FACTOR METHOD

The principle of this method is best explained by reference to a practicle example, for instance the analysis of binary alloys of lead and tin. Table 7.2 lists the count data obtained from a series of these alloys. The corresponding calibration curve is shown in fig. 7.3 from which it will be seen that this is a case of negative deviation. As is general, the reason for this may be due to

TABLE 7.2

Count rate data for tin/lead alloys

Sample No.	% Sn	c/s (Kα)	% Pb	c/s Lα
1	100.0	497.1	0	4.2
2	74.98	313.7	24.97	176.8
3	59.97	227.1	39.92	260.2
4	40.00	143.8	59.89	372.8
5	20.18	65.9	79.84	472.9
6	0	2.1	100.0	568.4

differences either in absorption or in enhancement. In this particular case, Sn Kα at 0.49 Å can scarcely excite the Pb Lα line (absorption edge at 0.95 Å) thus the effect must be explained in terms of differences in absorption. However, the absorption coefficient for Pb Lα is very similar in lead and in tin (134 for tin and 128 for lead, from the tables in appendix 2; or 126 and 117 respectively from International Tables[41]). This mass absorption coefficient would be expected to give, if anything, a positive deviation effect. However, as was pointed out in Section (1.4.4), under certain circumstances the absorption of the primary radiation may be more important than that of the secondary radiation (Equation 1.26).

The absorption coefficient for wavelengths shorter than the L_{III} edge of lead (0.95 Å) is very much larger for lead than for tin. For instance at 0.75 Å the approximate values are 40 for tin and 160 for lead. This in turn means, that the efficiency factor $C(\lambda\lambda_j)$ for every exciting wavelength, is higher at high tin concentrations than at high lead concentrations. Thus, as more lead is added, the slope factor m decreases to a minimum value of 5.684 c/s% at 100 % lead.

In setting up a mathematical correction formula, the assumption is made that the increase in the m value for lead is proportional to the concentration of tin. If this assumption is correct the following relationship will hold:

$$\text{concentration}_{Pb} = \frac{(R_p)_{Pb}}{m_{Pb}} [1 + a(\text{concentration})_{Sn}] \qquad (7.6)$$

where m_{Pb} is the counting rate per percent for pure lead and a the influence factor of tin on lead.

In terms of intensity:

$$(R_p) \text{ true}_{Pb} = (R_p) \text{ measured}_{Pb} [1 + a(\text{concentration})_{Sn}] \qquad (7.7)$$

The influence factor can be determined by substitution of calibration data from suitable standards. The average value of a from such a series of mea-

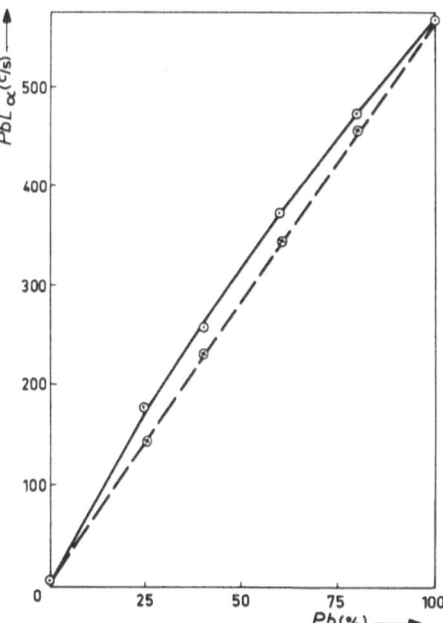

Fig. 7.3 Calibration curve for lead in a series of lead/tin alloys. The upper curve shows negative deviation due to increasing absorption of the primary radiation as more lead is added. This effect can be corrected for simply by applying the influence factor method described in Section 7-.5 (ii) (lower curve).

surements was found to be —0.0021 for the given case and using this correction factor, a reasonably linear correlation is obtained for all of the data listed in Table 7.2, although it will be seen by inspection that the largest deviations occur at low lead concentrations. Using this value of a, the concentration of lead in, for example sample 3 may be corrected to 40.0% compared to 45.7% for the uncorrected value which is obtained by rationing against the pure lead sample.

In very general terms the influence factor equation can be written

$$C_i = \frac{(R_p - R_b)i}{m_i} \cdot [1 + \Sigma C_j a_j] \qquad (7.8)$$

where C_i is the concentration of element i, R_p and R_b the peak and background counting rates corresponding to C_i, and m_i the counting rate per percent for pure element i. C_j and a_j are the concentration and influence factors for any element j. The sign of the influence factor is negative in the case of the correction of a curve showing negative deviation (i.e. μ interfering element $< \mu$ measured element) and positive for the more usual case of positive deviation (i.e. μ interfering element $> \mu$ measured element). Equation (7.8) also corrects for enhancement effects since these can be assumed to correspond to negative deviation.

The form of the influence factor Equation (7.8) has to be modified where ratio measurements are employed since as the reference sample, by definition, contains all of the elements of interest, the concentration term inside the matrix correction bracket becomes a difference term. Since an X-ray spectrometer performing a ratio measurement gives a number of counts N and an analysis time T rather than an actual count ratio, it is also necessary to replace the count rate term R by N/T.

A general form of the influence factor equation could thus be written:

A general expression for the solution of matrix effects

$$C_i^x = \left[\frac{\left(\dfrac{N_i^x}{N_i^s}\right) \cdot \left(\dfrac{T_i^s - N_i^s \cdot t}{T_i^x - N_i^x \cdot t}\right) - K_i^x}{1 - K_i^x}\right] \quad C_i^s \quad \left[1 + \sum_j k_{ij}\left(C_j^s - C_j^x\right)\right] \quad (7.9)$$

Concentration in sample	Intensity ratio	Dead time correction	Background correction	Concentration in standard	Matrix correction

The background correction is simplified somewhat since it is assumed that the background is constant over the selected calibration range (this is a reasonable approximation provided that line overlap problems are absent) and it will be seen that all data are corrected to a straight line defined by the ordinates K_i, 0 and I, C_i^s. One of the great advantages of the influence factor method is that it can be applied over a wide concentration range with exactly the same correction factors. One consequence of this, however, is that a wide count rate range is employed and in order to keep analysis times at low concentrations within reasonable limits, it is necessary to employ very high count rates at the high concentration range of the calibration. This in turn makes the use of a dead time correction obligatory either by means of a dead time correction term in the general equation as above, or by use of an automatic dead time correction circuit. As an example of the count rate ranges currently in use, in our own laboratory we work presently on a matrix correction procedure for the analysis of steels where we calibrate for chromium over the concentration range 0 to 30% and here the count rate range is from 0 to 170.000 c/s.

The slope factor method with and without certain subtle variation [50] has been applied routinely with great success to a variety of problems including copper based alloys, [34, 37] alloy steels [38, 40] and slags. [39] The method has the great advantage that once an influence factor has been determined for a certain set of instrumental conditions it should remain constant for all time. Thus by arranging for the count data output of the spectrometer to be in the form of punched tape the data can be fed straight into a suitable programmed computer allowing a very rapid and convenient analytical procedure.

A disadvantage of this influence factor method is the fact that the corrections are proportional to the unknown concentrations of the disturbing elements. However, the correction may be taken as proportional to the measured intensity of the disturbing element if these intensities are not influenced by changing concentrations of all the other elements. This leads to a much simpler set of equations. A large number of standards is necessary to determine the constants k_{ij}. However a certain relation should exist between these constants k_{ij} and the respective mass absorption coefficients. It has, indeed, been possible to calculate these constants starting from a simplified form of equation 1 - 23 and tabulated absorption coefficients. Only a few standards are necessary in this case, to determine the P_j's and to refine the calculated constants k_{ij}.

7.5.2 ABSORPTION CORRECTION METHODS

Many of the matrix interferences which are encountered in X-ray fluorescence spectrometry are due simply to variations in the mass absorption coefficients of the elements making up the matrix. This is certainly true of many alloys systems where by careful preparation of a specimen, surface effects can be easily avoided. Since reasonably accurate mass absorption coefficients are known for the majority of the wavelength range, the question arises as to whether or not absorption corrections can be applied directly. In principle, the total intensity Equation (1.23) should allow the possibility of calculating the intensities and of estimating the importance of variations in matrix absorption. This equation is however, rather unwieldy in the integral form as the integration has to be taken over the whole primary spectrum. If the secondary absorption term, $A \cdot C_i \cdot \mu_i(\lambda_j)$ dominates and the primary absorption term can be neglected, then the calculation becomes much simpler. This may be the case where the geometric factor A is very large, for example, at very low take-off angles.[42] If the primary absorption term cannot be neglected, it may be possible to replace the excitation by the total primary spectrum with that due to a hypothetical effective wavelength and to use the absorption coefficient for this wavelength, μ_e, to calculate the matrix absorption. Equation (1.23) reduces in this case to:

$$I_{js} = C_j I_{jp} \frac{\Sigma_i [C_i \mu_i(e) + A_i C_i \mu_i(\mu_j)]_p}{\Sigma_i [C_i \mu_i(e) + A_i C_i \mu_i(\mu_j)]_s} = C_j \cdot I_p \cdot \frac{\mu_p}{\mu_s} \qquad (7.10)$$

where I_{js} and I_{jp} are the respective intensities for the element j in the sample and as a pure element, C_j its concentration, and μ_p and μ_s the matrix absorption coefficients for the pure element and the sample.

Equation (7.10) can now be used to correct measured intensities. If the tin data from Table 7.2 are plotted as in Fig. 7.4 the calibration curve obtained is seen to exhibit strong positive deviation. The values of μ_s for the samples 1 to 6 may be calculated using μ_{Pb} for SnKα = 52; μ_{Sn} for SnKα = 12 and mass absorption coefficient values of 40 for tin and 25 for lead, referred to a

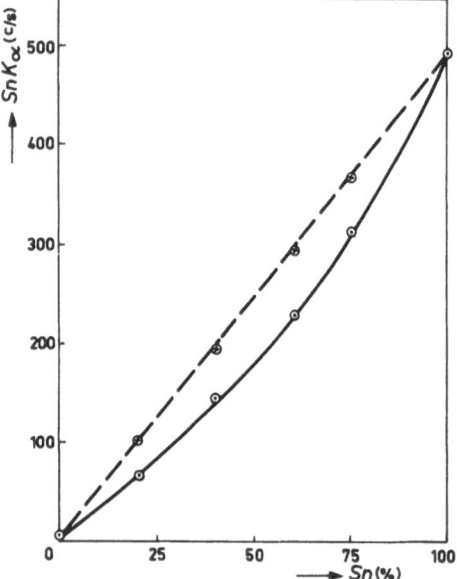

Fig. 7.4 Calibration curve for tin in a range of tin/lead alloys. The lower curve shows strong positive deviation due to large differences in the mass absorption coefficients of tin and lead for Sn Kα radiation. A linear correlation is, however, found after an absorption correction has been applied (see Section 7.5.2).

hypothetical effective wavelength, which is slightly shorter than the absorption edge of tin (0.43 Å). If the measured intensities of samples 2 to 5, as given in table 7.2, are now multiplied by the ratio of their respective matrix μ_s, to the matrix μ of sample 1, then values of 368, 292, 206 and 103 are obtained. If these values are replotted (Fig. 7.4) an almost linear calibration curve is obtained.

Unfortunately it is not possible to *calculate* the matrix μ before the actual composition of the sample is known, but very often this value can be measured directly by an independent experiment. Several workers[43-46] have described procedures for applying this method. One of these [44] involves the measurement of the fluorescent intensity from a pure specimen of the element to be analysed, but when placed behind the actual sample. Both the primary and the secondary radiation must pass the sample in exciting the pure element, thus, this procedure corrects for the total matrix μ. Sometimes this procedure can be simplified, particularly in those cases where the influence of the primary absorption can be neglected and the absorption of the characteristic wavelength only has to be considered. In this instance a weighed disc of the sample is prepared and the degree of attenuation by this disc, of monochromatic characteristic radiation emitted by the element to be analysed, is determined. This characteristic radiation can either be emitted by the sample or by a separate specimen containing a high concentration of the analysed element[43,47]. This method can be·used in a similar way to correct for absorption of the primary radiation, provided that it can be assumed that an effective wavelength exists and that its value can be established. As long as the usual precautions are taken to correct for equipment dead-time, this method can prove extremely useful for the analysis of elements giving short wavelength radiations (less than 2 Å or so), but it cannot be applied for longer wavelengths owing to difficulties involved in preparing a disc thin enough to allow a measurable quantity of radiation to be transmitted.

A further possibility is the correlation of the scattering power of the sample for coherent and/or incoherent radiation (cf. Section 7.2) with the matrix μ and using these data to correct for variations in the matrix absorption.[13,14]

REFERENCES

*　　1　Analytical Chemistry - Analytical Reviews (Bi-annual).
*　　2　BUWALDA, *Review of Literature*, XRFS 4th ed., Philips, Eindhoven, 1967.
　　　3　VOLBORTH, A., 1963, *Nevada Bureau of Mines Report 6*, part A.
　　　4　ADLER, I. and AXELROD, J. M., 1955, Spectrochim. Acta, **7**, 91.
　　　5　JONES, R. A., 1959, Analyt. Chem. **31**, 1341.
*　　6　GUNN, E. L., 1965, Appl. Spectr. **19**, 99.
　　　7　BERTIN, E. P., 1964, Analyt. Chem., **36**, 827.
　　　8　BACKERUD, L., unpublished.
　　　9　COMPTON and ALLISON, X-*rays in theory and experiment*, van Nostrand, New York, 1935, p. 191.
　　10　ANDERMAN, G. and KEMP, J. W., 1958, Analyt. Chem., **30**, 1306.
　　11　CULLEN, T. J., 1962, Analyt. Chem., **34**, 812.
*　12　KALMAN, Z. H. and HELLER, L., 1962, Analyt. Chem., **34**, 946.
*　13　REYNOLDS, R. C., 1963, Amer. Min., **48**, 1133.
　　14　CHAMPION, K. P., TAYLOR, J. C. and WHITTEM, R. N., 1966, Analyt. Chem. **38**, 109.
　　15　DWIGGINS, C. W., 1961, Analyt. Chem., **33**, 67.
　　16　TOUSSAINT, C. J. and VOS, G., 1964, Appl. Spectr. **18**, 171.
　　17　ROSE, H. J., ADLER, I. and FLANAGAN, F. J., 1962, U.S. Geol. Surv. Prof. Paper 450-B, 80.
　　18　WELDAY, E. E., BAIRD, A. K., MCINTYRE, D. B. and MADLEM, K. W., 1964, Amer. Min., **49**, 889.
　　19　RHODIN, T. N., 1955, Analyt, Chem., **27**, 1857.
　　20　SPIELBERG, N. and ABOWITZ, G., 1966, Analyt. Chem., **38**, 200.
　　21　WITMER, A. W. and ADDINK, N. W. H., 1965, Philips Serving Science and Industry, **12**, 2.

22 GUNN, B. L., 1961, Analyt. Chem., **33**, 921.
23 FELTEN, E. J., FANKUCHEN, I. and STEIGMAN, J., 1959, Analyt, Chem., **31**, 1771.
24 GILLAM, E. and HEAL, H. T., 1952, Brit. J. Appl. Phys., **3**, 352.
25 KOH, P. K. and CAUGHERTY, B. J., 1952, J. Appl. Phys. **23**, 427.
26 NOAKES, G. E., 1953, A.S.T.M. Spec. Techn. Publ. **157**, 57.
27 BEATTIE, H. J. and BRISSEY, R. M., 1954, Analyt. Chem., **26**, 980.
28 KEMP, J. W. and ANDERMAN, G., 1956, Spectrochim. Acta, **8**, 114.
29 CAMPBELL, W. J. and BROWN, J. D., 1964, Analyt, Chem., **36**, 312R.
* 30 SHERMAN, J., 1955, Spectrochim. Acta, **7**, 283.
31 SHERMAN, J., 1959, Spectrochim. Acta, **15**, 466.
32 ANDERMAN, G., 1966, Analyt. Chem., **38**, 82.
* 33 MITCHELL, B. J., 1958, Analyt. Chem., **30**, 1894.
* 34 LUCAS-TOOTH, H. J. and PRICE, B. J., 1961, Metallurgia, **54**, 149.
35 LUCAS-TOOTH and PYNE, *Advances in X-Ray Analysis*, Plenum, New York, 1963, **7**, 523.
36 ALLEY, B. J. and MYERS, R. H., 1965, Analyt. Chem., **37**, 1685.
37 BAREHAM, F. R. and FOX. J, G. M., 1960, J. Inst. Metals, 88, 344.
38 MARTI, W., 1962, Spectrochim. Acta, **18**, 1499.
39 JOHNSON, W., *Proceedings of the 4th M.E.L. Conference on X-ray Analysis*, (Sheffield, 1964), Philips, Eindhoven, 73.
40 LOUNAMAA, N. and FUGMANN, W., 1966, Jerkout. Ann., **150**, 99.
41 *International Tables for X-Ray Crystallography*, Vol. III. Kynoch Press, Birmingham, 1962.
42 EBEL, H., 1966, Z. f. Metallkunde, **57**, 454.
43 NORRISH, *Conference X-Ray Spectrographic Analysis*, Sydney, August, 1964.
* 44 CARR-BRION, K. G., 1965, Analyst, **90**, 9.
45 SMAGUNOVA, A. N., LOSEV, N. F. and LIPSKAYA, V. I., 1965, Industr. Lab., **31**, 201.
46 LEROUX, J. and MAHMUD, M., 1966, Analyt. Chem., **38**, 76.
47 GOLDMAN, M. and ANDERSON, R. P., 1965, Analyt. Chem., **38**, 109.
48 JOHNSON, W., 1967, Paper presented at 20th B.I.S.R.A. Chemists Conference, Scarborough.
49 JENKINS, R., 1968, Philips Scientific and Analytical Equipment Bulletin, 79.177/FS 15, Philips, Eindhoven.
* 50 LACHANCE, G. R., and TRAILL, R. J., 1966, Canadian Spectroscopy, **11**, 43.

CHAPTER 8

SAMPLE PREPARATION

8.1 General

Since X-ray spectrometry is essentially a comparative method of analysis, it is vital that all standards and unknowns be presented to the spectrometer in a reproducible and identical manner. Any method of sample preparation must give specimens which are reproducible and which, for a certain calibration range, have similar physical properties including mass absorption coefficient, density and particle size. In addition the sample preparation method must be rapid and cheap and must not introduce extra significant systematic errors, for example, the introduction of trace elements from contaminants in a diluent.

Since sample preparation is an all important factor in the ultimate accuracy of any X-ray determination, many papers have been published describing a multitude of methods and recipes for sample handling. As several excellent review articles are available in the literature[1-5] the contents of this chapter will be restricted to a general summary of the problem and only the more important sample preparation techniques will be discussed.

In general samples fit into three main categories:

1) Samples which can be handled directly following some simple pre-treatment such as pelletizing or surfacing. For example, homogeneous samples of powders, bulk metals or liquids (Section 8.2).
2) Samples which require significant pre-treatment. For example, heterogeneous samples, samples requiring matrix dilution to overcome inter-element effects and samples exhibiting particle size effects (Section 8.3).
3) Samples which require special handling treatment. For example, samples of limited size, samples requiring concentration or prior separation and radioactive samples (Section 8.4).

Each of these categories will be discussed in turn and where possible practical examples will be given illustrating the various sample treatments which are available.

8.2 Samples requiring only a simple treatment

8.2.1 BULK SOLIDS (a) Metals

Samples may be submitted for analysis in a variety of shapes and sizes and for homogeneous specimens two problems are of particular importance in presenting them to the spectrometer. First, it is necessary that all samples and standards cover the same irradiated area of the sample cup and second, all samples and standards must have a surface finish which is of sufficient fineness and free from surface contamination. In routine analysis the first of these problems can be avoided by arranging for samples to be submitted in the form of cast circular billets of sufficient diameter to completely fill a standard sample cup. Specimens of suitable thickness can be taken by slicing the billets with a diamond cut-off wheel. In the case of awkwardly shaped samples such as wires or small irregularly shaped sections it may be impracticable to cut a specimen which can be handled directly. This is certainly true of samples which are submitted in the form of drillings or filings and here the simplest handling method is to take the sample into solution since this provides the most convenient way of allowing comparisons to be made with standards. The standards can in turn be prepared from suitable soluble metal salts. Where solution techniques have to be rejected on the grounds of time or because samples cannot be destroyed, useful but less accurate data can frequently be obtained either by masking down the X-ray beam such that a reproducible area from each specimen is obtained,[6] or by comparing the intensities of each analytical line as an indication of relative concentrations.[7] Under certain conditions, e.g. where the material is soft, high pressure pelletizing the drillings or filings can give analysis samples of sufficient reproducibility.

The problem of giving all samples a surface finish of sufficient fineness is really twofold since it is necessary that the degree of fineness be defined with some certainty before attempts are made to obtain this finish. In the section on matrix effects some mention was made of intensity variations due to local shielding of individual spheres and, although this argument was introduced with specific reference to particle size variations in powders, a similar situation can occur in bulk metals. Here one has to consider local shielding of neighbouring maxima on a nominally flat sample surface. The shielding effect is illustrated in Fig. 8.1. The sample surface is indicated by the full line and primary radiation entering at angle ψ_1 penetrates to an effective depth shown by the dotted line. Any greater depth than this can be considered as infinite as far as the measured wavelength is concerned, hence only the non-hatched

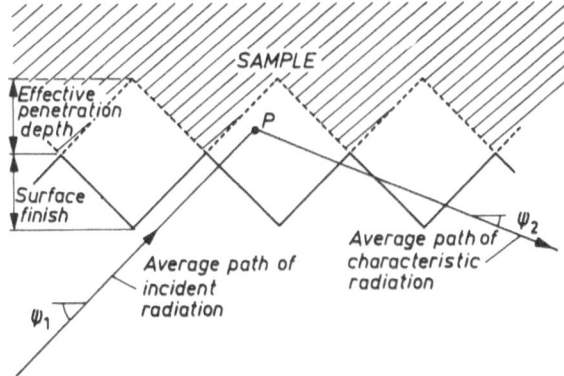

Fig. 8.1 Effect of shielding due to surface finish.

area can contribute to the measured intensity. The effective depth can be estimated from the take-off angle ψ_2 of the spectrometer and the path length x of the measured radiation. For major concentrations the absorption co-efficient for the primary radiation is greater than that for the secondary radiation and in this case the effective depth is determined by both primary and secondary absorption. In the general case x can be calculated from the standard absorption equation, $I = I_0 \exp\left(- \mu\rho x\right)$ and for 99 % absorption x is approximately equal to $4.61/\mu\rho$, where μ is the mass absorption coefficient of the matrix for the measured wavelength and ρ is the matrix density. Radiation arising at point P, although theoretically within the effective penetration depth, cannot escape from the matrix because of the additional path length introduced by the interposed maximum on the sample surface. The measured intensity is therefore lower than would be expected. Since the degree of shielding varies with the path length of the measured radiation in the sample matrix, the required surface finish will be dependent both upon the wavelength of the measured radiation and the composition of the sample matrix. In general the correlation between X-ray intensity and surface finish is similar to that shown in Fig. 8.2. The majority of elements giving wave-lengths shorter than about 2 Å exhibit the correlation shown by curves A and B and it is useful to define the point of deviation of curve B from linearity, $(S)_{max}$, since this represents the coarsest surface finish that can be employed without significantly decreasing the measured intensity. Where the approximate composition of the matrix is known the optimum surface finish can frequently be predicted since $(S)_{max}$ is roughly proportional to the cube root of the path length x.[8] In general, the longer the wavelength of the measured element and/or the larger the mass absorption coefficient of the matrix, the more critical will be the required surface finish. The loss in intensity due to surface roughness is in itself not serious, but it may influence the comparison between sample and standard as the surface effects should be the same in both cases. Table 8.1 illustrates some values of required surface finish for typical matrices.

The best method of obtaining the optimum surface finish depends to a very large extent upon the physical properties of the sample. For moderately hard metals a variety of methods have been successfully employed including finishing, surface grinding, lapping or turning with a capstan lathe or an end miller. However, machining methods are of little use for the surfacing

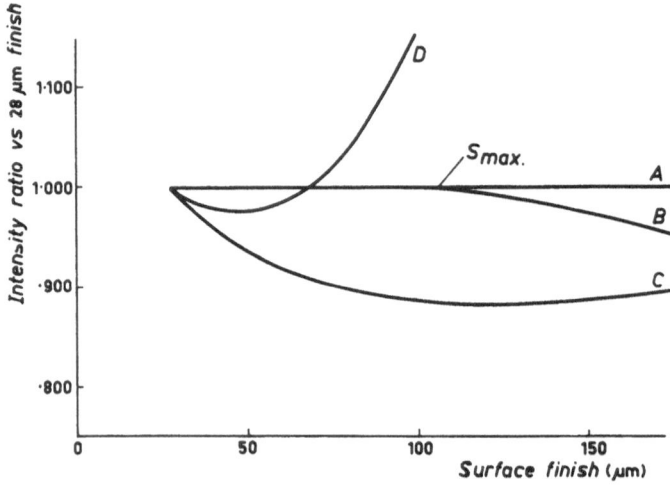

Fig. 8.2 Variation of X-ray intensity with surface finish.

of very hard materials and here lapping is probably the most efficient of the currently available methods.

Special problems due to smearing arise in the preparation of very soft materials and particularly where one constituent of a matrix is softer than the remainder. Curves C and D in Fig. 8.2 are typical of systems where smearing has taken place. Since smearing is essentially due to small scale heterogeneity the problem will be considered later in the preparation of inhomogeneous samples.

TABLE 8.1

Surface finish requirements for typical alloy systems

Sample	Element	Concentration	Analysis line	Path length (μm)	S_{max} (μm)
Aluminium/	Al	92.59	Kα	23	70
Copper	Si	0.16	Kα	25	80
	Fe	0.27	Kα	203	>180
	Cu	4.43	Kα	48	>·180
Gun metal	Cu	87.38	Kα	53	130
	Zn	3.75	Kα	64	120
	Sn	5.25	Kα	288	>180
			Lα	8.4	60
	Pb	3.60	Lα	27	125
			Mα	3.3	60
High Alloy	Mn	0.55	Kα	63	>180
Steel	Ni	2.87	Kα	17	>·180
	Cr	20.40	Kα	51	130
	Fe	62.16	Kα	39	140

Whatever method of surfacing is employed special care must be taken to ensure that no extra surface contamination is introduced. This is particularly important in the case of softer metals and alloys where particles of grinding agent may become embedded in the sample surface during treatment. All traces of lubricant, cutting fluid, finger marks and so on must be removed by use of a suitable solvent immediately prior to examination. Extra precautions should be taken whenever samples are prepared some time before analysis, particularly when the atmosphere is known to contain significant quantities of moisture, hydrogen sulphide or sulphur dioxide, the presence of which can lead to the formation of surface films of oxides or sulphides. Once the sample has been prepared care should be taken to ensure that it fits properly into its sample cup since displacement of as little as 100 microns can make a difference of as much as 0.5% in count rate. Wherever possible, samples should be rotated in their own plane during analysis since this helps to even out surface irregularities and, in addition, gives a larger integrated area of analysis.

8.2.2 BULK SOLIDS (b) Non-metals

Most of what has already been said about a preparation of metals applies similarly to non-metals. Samples of suitable size can usually be cut from finished stock including glass, ceramic or plastic materials by use of a diamond trepanning tools or punches. Pieces of fibre or cloth can usually be clamped in a cell between two sheets of polyethylene terephthalate (mylar) film or alternatively mixed with a resin bonded polymer and cast into blocks.

8.2.3 POWDERS

Where powders are not affected by particle size limitations the quickest and simplest method of preparation is to press them directly into pellets of constant density, with or without the additional use of a binder. In general, provided that the powder particles are less than about 50 microns in diameter (300 mesh) the sample will pelletize at around 2-5 tons per square inch. Where the self-bonding properties of the powder are poor, higher pressures of perhaps up to 50 tons per square inch may have to be employed or in extreme cases recourse made to the use of a binder. High pressure pelletizing in a die, or directly in a sample cup often results in a fracture of a pellet following the removal of pressure from the die. The fracturing is due to slight deformation of the die under pressure and can be avoided by pressing the powder in a former which will irreversibly yield under the high pressure. For example, one method of sample preparation which has been employed successfully in the analysis of slags and sinters consists of pressing the sample in lead rings or rubber "O" rings between two hardened steel plattens. [9]

It is sometimes necessary to add a binder before pelletizing and the choice of the binding agent must be made with care. As well as having good self-bonding properties the binder must be free from significant contaminant elements and must have low absorption, (unless for some reason the mass absorption coefficient of the matrix needs to be increased). It must also be stable under vacuum and irradiation conditions and it must not itself introduce significant interelement interferences. Of the large number of binding agents which have been successfully employed probably the most useful are starch, ethyl cellulose, lucite, polyvinyl alcohol and urea. Use of a binding agent invariably decreases the overall matrix absorption and so dilution of a sample with a certain quantity of binder does not necessarily mean that the sensitivity for a certain element will drop by an equivalent amount. Frequently the addition of one or two parts by volume of a carefully chosen binder makes little or no difference to the absorption of a medium average atomic number matrix. However, the addition of the binder can lower the average atomic number of a sample quite significantly with a subsequent increase in the scattering power of the sample. Thus addition of a binder often increases the background radiation of a sample and this can become important in the determination of small quantities of elements with wavelengths shorter than about 1 Å where background is difficult to remove. As a result of this it is often best to use the minimum quantity of binder mixed with the sample and to add additional binder as a backing for the sample to give it extra strength. This backing technique can also successfully be employed when the available amount of sample is too small to give a pellet of sufficient mechanical strength. Where necessary, further stability can be achieved by spraying the pellet with a one per cent solution of formvar in chloroform.

8.2.4 LIQUIDS

Provided that the liquid sample to be analysed is single phase and relatively involatile, it represents an ideal form for presentation to the X-ray spectrometer. Where the vapour pressure of the liquid to be analysed is too great to allow the use of a vacuum path a special sample cup or a hydrogen or helium path instrument must be used for the measurement of the longer wavelengths. (e.g. greater than 3 Å). The liquid phase is particularly convenient since it offers a very simple means for the preparation of standards. Since most matrix interferences can be successfully overcome by taking the sample into liquid solution, this technique represents the final stage of many sample handling methods.[2,10,11] Although the majority of matrix interferences can be removed by the solution technique, the process of dealing with a liquid rather than a solid can itself present special problems which in certain instances can limit the usefulness of the technique. For example, the taking of a substance into solution inevitably means dilution and this, combined with the need for a support window in the sample cell (unless special optics are employed), plus the extra background arising from scatter by the low atomic number matrix, invariably leads to a loss of sensitivity, particularly for the longer wavelengths (greater than 2.5 Å).

Problems can also arise from variations in the thickness and/or composition of the sample support film. Fig. 8.3 compares the attenuation of wavelengths in the 3-10 Å region by different types of support film. The most commonly used type of film is 6 μm polyethylene terephthalate and it will be seen that intensities of the longer wavelengths are considerably reduced. It is possible to reduce the thickness of polyethylene terephthalate film down to about 3 μm before pin-hole formation becomes a problem and use of this thinner film somewhat reduces the total attenuation. One apparently insurmountable problem with polyethylene terephthalate films is the presence of large inclusions of high absorption, consisting of, among other

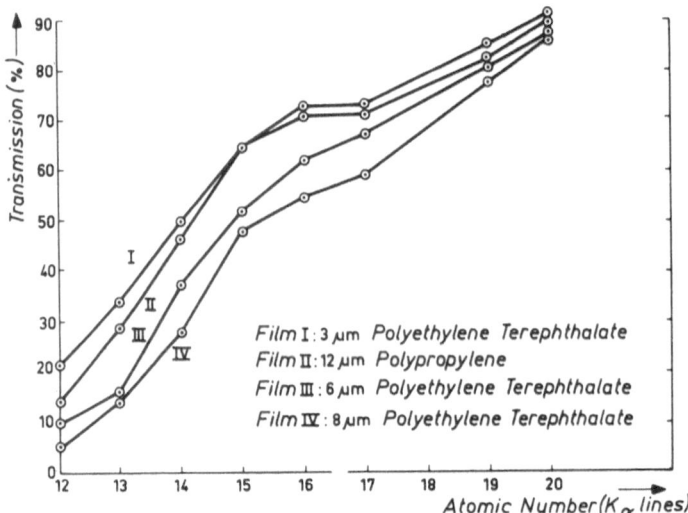

Fig. 8.3 Transmission of support films.

things, antimony salts from the copolymer catalyst used in the manufact-
uring process. Although the total concentration of impurity is relatively
small (typical ash content 0.2 weight %) the average atomic number of the
inclusions is high and anomalous results can occur when measuring wavelengths greater than about 6 Å. An alternative support film is 12 μm oriented
polypropylene which is attractive since its ash content is low (usually less
than 0.005 weight %) and impurities are relatively evenly distributed. Also
since its average atomic number is less than one half that of polyethylene
terephthalate its absorption coefficient is lower by a factor of about five.
Thus even allowing for the need for a 12 μm film thickness the attenuation by
polypropylene film is a factor of two less than 6 μm polyethylene terephthalate. The greater film thickness is required to overcome the tendency of the
oriented polypropylene to sag under the weight of the sample. Care should
be taken to select only fully oriented polypropylene for cell windows since
even partially oriented material may sag. Non-oriented polypropylene
(such as is used for making thin flow counter windows) must be avoided at
all costs!

The process of taking a sample into solution can be rather tedious and
difficulties sometimes arise where a substance tends to precipitate during
analysis. This itself may be due to the limited solubility of the compound
or to the photochemical action of the X-rays causing decomposition. In
addition systematic variations in intensity can frequently be traced to the

formation of air bubbles on the cell windows following the local heating of the sample. Despite these problems, the liquid solution technique represents a very versatile method of sample handling in that it can remove nearly all matrix effects to the extent that accuracies obtainable with solution methods approach very closely the ultimate precision of any particular X-ray spectrometer. Some care must be taken to match the mass absorption coefficient and density of standard and sample very closely, and there is also some evidence to suggest that variations on the pH of a solution may cause significant systematic errors. Density variations can be particularly troublesome, especially in the analysis of hydrocarbon solutions. For example, when analysing for heavy metals such as lead, differences of about 15-20 % are found in the relative intensities of identical concentrations of organometallic lead in benzene and cyclohexane. Fortunately, it is often possible to use coherently or incoherently scattered tube lines as internal standards to correct for both density and mass absorption coefficient variations.[12]

Where liquids of more than one phase are to be analysed, prior separation must be made before specimens are presented to the spectrometer. A typical example is the analysis of used lubricating oils where, in addition to the hydrocarbon matrix containing unused additives, decomposition products and fine metal particles may also be present. In this instance separation can be brought about by dilution with a light solvent followed by centrifuging out the solid matter. The centrifugiate can then be treated as a heterogeneous solid, and the supernatent liquid treated as a homogeneous solution. If necessary, the light solvent can be distilled off.

8.3 Samples requiring significant pre-treatment

8.3.1 BULK SOLIDS

Bulk solids requiring significant pre-treatment before analysis generally fall into two categories (i) heterogeneous samples and (ii) homogeneous samples which require dilution in order to mask interelement effects. Both types of sample can be successfully handled by taking into solution and indeed this is invariably the only method of dealing with heterogeneous bulk samples. The solution technique employed need not necessarily involve a liquid since the advantages to be gained by liquid solutions can frequently be obtained just as satisfactorily by use of solid solutions. Care should be taken not to reject the solid solution technique out of hand simply on the grounds of apparent inertness of the sample to the solid solution melt, since

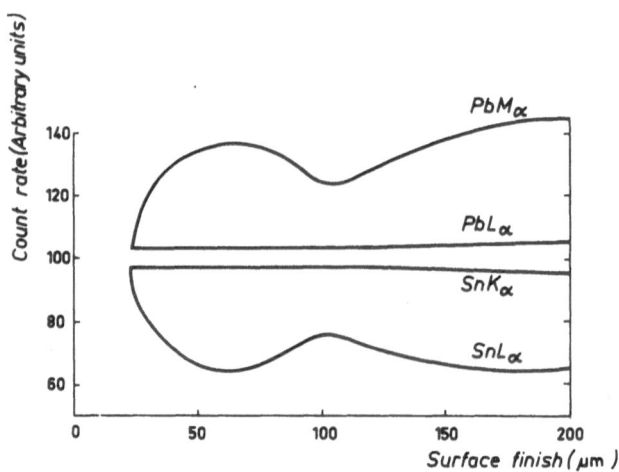

Fig. 8.4 Effect of surface finish on a 50% Tin/50% Lead Alloy.

this problem can often be overcome by recourse to a fairly simple pretreatment. For example, metal filings can be taken into solid solution with borax following pretreatment with flowers of sulphur to convert to sulphides.[13]

The success with which heterogeneous samples can be handled directly depends to a large extent on the scale of the inhomogeneity. If this is on the micro scale, as is the case with most metals, the only real problem is one of surfacing. Should relatively large scale inhomogeneity be present in the sample then again solution techniques have to be used.

Surfacing problems in a sample showing microscale inhomogeneity often arise from differences in the ease with which hard and soft components in the sample surface can be cut. Since a surfacing technique must of necessity be sufficient for the hardest matrix constituent it is far more than sufficient for the soften components. The result is that softer particles in the sample are dragged out and smeared over the prepared surface, resulting in an increase in intensity from lines due to the softer constituent and decrease in the intensity for all other lines. The effect of smearing on X-ray intensity becomes worse as the wavelength of the measured element becomes longer. The result is demonstrated in Fig. 8.4 which shows the effect of surface smearing of lead on the intensity of tin and lead lines in a 50/50 lead/tin alloy. When the relatively long wavelength lines i.e. lead Mα and tin Lα are used, the effect is of the order of 20% at 100 microns, but this can be minimised by the use of the shorter wavelength lines i.e. lead Lα and tin Kα. The decrease of the smearing effect on the intensity when shorter wavelengths are

used is due simply to the relatively longer path lengths of the harder radiations. For instance, in this matrix lead Lα has a path length of 27 μm compared to only 3.3 μm in the case of lead Mα. This dependency of the intensity
ratio of difference series wavelengths from the same element can be used to
judge whether or not a surface problem exists in routinely prepared samples.
Other cases occur where surface effects cannot be minimised so easily, typical
examples being lead in high leaded brasses and aluminium in aluminium
silicon alloys. Careful finishing on 5-10 μm diamond dust paste helps to
minimise the effect, but is to be hoped that the newer techniques of sample
handling, such as spark planing, may avoid the problem completely. [8]

8.3.2 POWDERS

The analysis of powders is invariably more complex than that of bulk
metals, since the addition to interelement interferences and macroscale
heterogeneity, particle size effects are also important. Although inhomogeneity and particle size can often be minimised by grinding and pelletizing
at high pressure, often the effects cannot be completely removed because the
harder compounds present in a particular matrix are not broken down.
The effect is to produce systematic errors in the analysis of specific types of
material typified by silicious compounds in slags, sinters and certain minerals. The best way of completely removing these effects is to employ a fusion
technique based on the method of Claisse.[14,15] The original Claisse technique consisted of fusing the sample with borax and casting into a solid
button, but many variations of the original method have been described,[16]
of which the more important are the use of lithium tetraborate[17] as an
alternative to the sodium salt. Lithium tetraborate offers the advantage of
having a lower average atomic number than the corresponding sodium salt
and if used with lithium carbonate, with which it forms a eutectic mixture,
when in the ratio of 6:1, it has a melting point lower than that of borax. It is
however, somewhat hygroscopic and prefused mixtures of lithium borate —
lithium carbonate should be stored in well stoppered bottles. This point
should also be remembered when working with fused, ground and pressed
discs, since the pick up of moisture is quite a rapid process. X-ray diffraction
measurements have indicated the presence of Li_2CO_3, H_2O on the disc
surface after they have been left for as little as 24 hours in a humid laboratory. The actual fusion reaction can be performed at 800-1000 °C containing the fusion mixture in a crucible made of, for example, platinum,
nickel or silica. Each of these materials has certain advantages to offer but all suffer from the disadvantage that the melt tends to wet the
sides of the dish and it is impossible to effect a complete recovery of the

fused mixture. Use of graphite crucibles overcomes this to some extent but probably the neatest way of avoiding the difficulty is to use a crucible of platinum + 3% gold. This alloy is hardly wetted at all by borate fusion mixtures, wastage is avoided and cleaning is made much easier. Although the source of heat is usually a standard laboratory muffle furnace or even simply a Meker burner, slightly more sophisticated methods have been tried of which the most promising is the high frequency heating furnace. Using such a device fusion plus reaction times of less than three minutes have been achieved, utilizing a fully automatic process. [41]

All of the bead techniques suffer from the limitation that the necessary annealing process may take several hours to complete although this time can often be reduced by preparing very thin beads,[18] or by special handling of the bead.[19] The time lost can, however, be almost completely avoided by grinding the bead immediately after fusion and pelletizing the resultant powder.[20] Other fusion techniques employ sodium carbonate,[21] sodium and potassium bisulphates[22] with or without added sodium fluoride[23] and ammonium metaphosphate[24] as the reacting melts, followed by pelletizing the melt or dissolving in water and analysing the resultant solution. A glass, formed by fusion with borax, may partially recrystallize during the cooling process and it is necessary that any microscale heterogeneities be avoided. This may be checked by taking an X-ray diffraction pattern of the powdered bead where no diffraction lines should be observed.

Many of the problems and failures encountered in the preparation of fused beads are due to the lack of appreciation of the chemical processes which occur during the fusion process. Essentially the purpose of the fusion is twofold, first to completely react the compounds present in the sample with the fusion mixture to give total solution and second to cool the fused melt to form a solid glass (i.e. completely non-crystalline). Incompleteness of either of these two stages will result in an inhomogeneous sample which, in addition to being bad in itself, can also lead to cracking of the glass bead during the cooling stage. For example, fusion with sodium or lithium borates will yield glassy borates of the elements in the sample but the number of stages in the reaction may be considerable. For instance a divalent metal oxide MO may produce any number of reaction products:

a. $Na_2B_4O_7$ $\longrightarrow$ $2NaBO_2 + B_2O_3$
    sodium tetraborate   sodium metaborate

b. $NaBO_2 + MO$ $\longrightarrow$ $NaMBO_3$
                          orthoborate

c. $B_2O_3 + MO$ $\longrightarrow$ $M(BO_2)_2$
                          metal metaborate

d. $M(BO_2)_2 + 2NaBO_2 \longrightarrow Na_2 M(BO_2)_2$
complex ·borate

The final distribution of reaction products depends to a large extent on reaction temperature and initial sample to borax weight ratio. It is, however, important to avoid reaction products which lie outside the glassy region of the system and such problems can often be avoided simply be reference to the appropriate phase diagram — many of which are available in the literature particularly for the lithium borate system [43]. Movement to the appropriate part of the phase diagram can always be achieved quite simply by varying the initial weight ratio of sample/borax/boric acid thus achieving some control over the reaction products. As an example, in the series of reactions indicated above the production of metal orthoborate (reaction c) can be made the preferred reaction by ensuring an excess of B_2O_3 in the reaction mixture, in practice this would be done by using a mixture of borax and boric acid as the reaction mixture rather than borax alone.

It is also important to realise that sodium tetraborate is the weak acid salt of a strong base and as such may be rather unreactive towards very basic materials. Lithium borate being the weak acid salt of a weak base may be far more suitable for this type of sample. Where reaction proceeds rather slowly the process may be accelerated by use of oxidizing conditions formed by use of oxidizing agents such as potassium nitrate or barium peroxide in the reaction mixture. Again it should be remembered that use of graphite crucibles always tends to give reducing conditions and may be rather unsuitable for certain types of material such as easily reducible oxides.

As with liquid solutions all of the solid techniques suffer from the fundamental disadvantage of high dilution and the increase of scattered background. The dilution required to overcome a matrix effect can sometimes be minimised by use of smaller quantities of fusion mixture along with high absorbing compounds such as lanthanum oxide. Matrix absorption can thus be stabilised for standards and samples without significantly increasing the scattering power of the matrix. A great advantage of the solid solution method over the liquid solution method is that cell windows are not required and vacuum conditions can still be employed.

8.4 Samples requiring special handling treatment

8.4.1 VERY SMALL SAMPLES

Where only very limited quantities of material are available for analysis, considerable ingenuity may be required on the part of the analyst in reproducibly presenting a suitable specimen to the spectrometer in a manner which will provide the maximum X-ray intensity from the elements to be analysed. A major problem in the quantitative analysis of small quantities is that there is great difficulty in accurately weighing a sample whose total weight is only of the order of a few milligrams. Where a sample weight can be obtained with a sufficient accuracy, fusion techniques can be applied with some success particularly where the elements to be estimated are of relatively high atomic number. In the case of medium atomic number elements sufficient sensitivity can still be obtained even after very high dilutions, perhaps up to 1 : 10,000.[25] Probably a far easier way of handling weighable quantities of sample is to dissolve in a suitable solvent and evaporate onto filter paper discs. Care should be taken that always the same surface area be wetted; this can be achieved by applying a wax ring of fixed diameter on the paper and by using a fixed volume of solution.[26] This method is certainly potentially more accurate since internal standards can be added very precisely to the discs by spotting known amounts of dilute standard solutions with the aid of micro-pipettes.

There is in fact a very great incentive in spreading the sample thinly since the relative efficiency of production of useful X-rays falls off very rapidly with the increase in depth of the sample. This effect can be demonstrated with reference to Fig. 8.5a. If a small crystal of lead sulphate weighing about

a

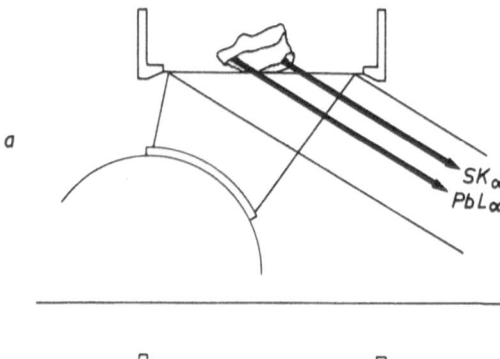

b

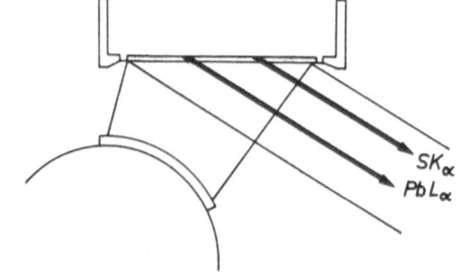

Fig. 8.5 Sample presentation. Since the path length of SKα is relatively short only a fraction of the lump of sample can contribute to the useful radiation (8.5a). However, by spreading the same amount of sample very thinly (8.5b) the whole of the sample can contribute ot the measured intensity.

10 milligrams were placed directly in a sample cup and irradiated it would be found that, whereas only a very small fraction of the total sample would effectively produce SKα radiation, practically the whole sample would contribute to the production of PbLα radiation. This is simply due to the differences in path lengths of SKα and PbLα radiation in PbSO₄. PbLα with a mass absorption coefficient of about 100 has a path length of approximately 70 μm. However as the absorption of the primary exciting radiation cannot be neglected in this case and since the take-off angle is approximately 35° in most spectrometers, only some 20 μm of the total thickness, contributes to the production of PbLα radiation which leaves the sample. SKα however, with a mass absorption coefficient of over 1000 has a path length of only 6 μm, hence only the radiation produced in the first few microns of the total depth of the sample can escape. By spreading the sample very thinly Fig. 8.5b (an average depth of about 2 μm can be obtained for a sample area of 6 cm²) all of the sample would now effectively contribute to both SKα and PbLα.

A further incentive for spreading a limited sample into a thin film and rotating it over the primary beam is that the actual positioning of a single particle of powder in a sample cup is critical both from the point of view of intensity and goniometer setting. The first of these points arises because the distribution of X-radiation over the irradiated area of the cup is far from uniform. Fig. 8.6 illustrates the reason for the variation in X-ray intensity

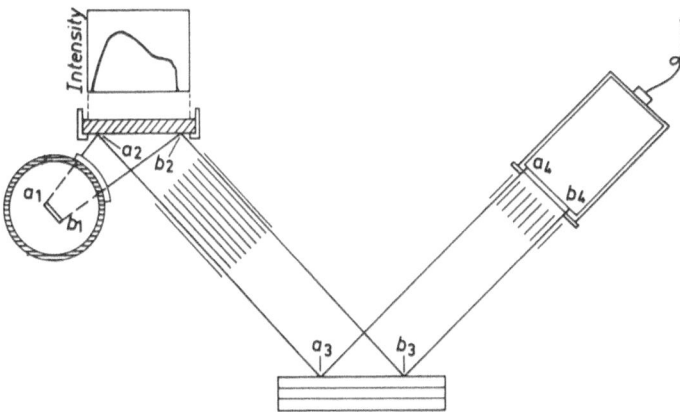

Fig. 8.6 Variation in intensity across the surface of a specimen. Owing to the fact that the distance $a_1 - a_2$ is significantly shorter than $b_1 - b_2$ the intensity distribution (see insert) falls off going from a_2 to b_2. It will also be seen that the majority of the radiation arriving at a_4 starts from a_2 and that from b_2 finishes up at b_4. This can give rise to peak shift effects when a specimen is used which is significantly smaller than the area defined by the mask of the X-ray tube window.

between the extreme edges of the sample cup. The distance between a_1 and a_2 is much less than that between b_1 and b_2 and since the X-ray intensity falls off as a complex function of the distance, the intensity at b_2 is far less than that at a_2. The small insert gives an indication of the variation in intensity between the extreme edges of the cup. The effect can be further complicated by the fact that different areas of the analysing crystal diffract radiation from different portions of the sample. Both of these effects can be completely removed by rotating the sample during analysis. The effect of the position of a small sample on the maximum angle of diffraction can also be seen from Fig. 8.6. Radiation arising at a_2 will be diffracted at a_3 and enter the detector at a_4. Similarly, radiation arising at b_2 will enter the detector at b_4. If the sample area is small in comparison with the irradiated area of the sample cup, movement of the sample from b_2 to a_2 will have the effect of increasing the angle of maximum diffraction. The relatively large solid angle of acceptance of the detector helps to minimise this effect but angular variations of up to 0.5° 2 θ are still possible.

The spreading of the sample into a thin film represents an extremely attractive means of analysis for a number of other reasons not the least of which is that matrix effects are often negligible (see Chapter 7). The limit as far as sample preparation is concerned is that the relative intensity of an analytical line is related both to the total weight and the self absorption of the sample. One possible means of correcting for this is to measure the attenuation by the sample of an intensity from a sample giving a similar wavelength to that being studies. This method is particularly useful where trace quantities of an element are being measured in a major constituent whose concentration can be assumed to be constant.

A good example of this is to be found in the analysis of small (about 5 milligrams) samples of pure calcium carbonate for traces of strontium (for example in the analysis of shells from molluscs).[27] Where large quantities of sample are available, the ratio of the count rate for strontium, to that for calcium can be used as a measure of the concentration of the strontium. The calcium can of course be assumed to remain constant even for variations of up to a few per cent of strontium. For small samples, however, the most sensitive method of sample preparation is to spread the sample over the surface of a piece of scotch tape. Although ratioing the strontium and calcium count rates should at first sight correct for variations in the volume of sample and standards irradiated, in practice this does not work. As will be seen from Fig. 8.7 the absorption of the sample is very different for calcium and strontium over the range of sample thickness employed and the ratio method over-corrects for sample volume. By measuring the attenuation by the sample

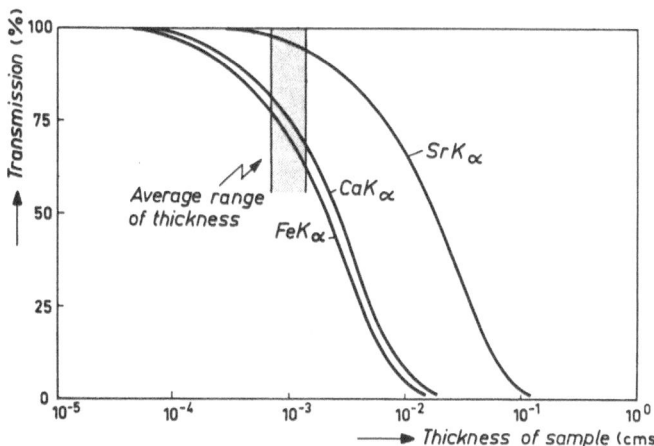

Fig. 8.7 Variations in transmission efficiency of thin sample of Calcium Carbonate for the Kα radiations of Strontium, Calcium and Iron.

of radiation from an iron disc, however, a better means of correction is found since the absorption of the matrix for calcium Kα radiation is almost identical to that for iron Kα radiation. In this instance a linear correlation is found between concentration of strontium and the function (Count rate due to iron passing through the sample) × (Peak minus background count rate due to strontium).

Thin film methods have also been employed with considerable success in the analysis of surface deposits [28,29)] and segregations.[30] For example, surface deposits on bulk metals can frequently be removed simply by use of scotch tape which can then be removed complete with the adhering layer and treated directly as any other thin film.

8.4.2 VERY DILUTE SAMPLES

Where the concentration of an element in a sample is too low to allow analysis by one of the methods already described, work-up techniques have to be employed in order to bring the concentration within the detection range of the spectrometer. Concentration methods can be employed where sufficiently large quantities of sample are available and the efficiency of the work-up technique depends to a very large extent on the ingenuity of the operator. Techniques which have been employed are necessarily rather diverse in character. For example, gases which are contaminated with solid particles can be treated very simply by drawing the gas through a filter disc followed by direct analysis of the disc.[31)] The technique has also been successfully employed for the analysis of trace amounts of zinc and lead in air. In this instance actual amounts of less than a microgram could still be detected on the discs with a normal spectrometer. Low concentrations in solution can be concentrated with ion exchange resins[32] and eluted and evaporated onto filter paper discs. Alternatively the ion exchange resin itself can be used as a sample support either by pelletizing [33-35] the exchange resin or by use of ion exchange filter papers.[36,37] This latter method has proven to be extremely useful for the determination of metals in very dilute solutions since it is possible to select ion exchange resins which are reasonably specific for one type of ion. Concentration is easily brought about by shaking the solution with a known amount of resin for several minutes, followed by filtering, drying and pelletizing. For example, 0.1 ppm of gold in solution can be determined by this method with a total sample preparation time of less than 15 minutes. Concentration can sometimes be effected simply by evaporating the solution straight onto confined spot filter paper,[38-40] and this technique has been employed with great success in the field of clinical chemistry. An alternative method which has also given excellent results in this field[41] entails direct evaporation of the solution onto planchettes.

8.4.3 RADIOACTIVE SAMPLES

The handling and pretreatment of radioactive samples presents two major problems additional to those already discussed. First the containment of a sample is all important since the potential health hazard can be considerable and it is vital that all of the precautions normally associated with the handling of active materials be strictly adhered to. Secondly, all radiation which can enter the detector, other than that arising as deliberately excited X-radiation from the sample, must be kept at an absolute minimum. Although the majority of published work on X-ray analysis of radioactive materials deals with diffraction rather than spectrometry, the problems of containment and shielding are similar. However, since it is invariably necessary that procedures based on X-ray spectrometry be rapid, in order to complete with alternative methods of analysis, the extra effort involved in preparing samples in suitable containers, combined with the relatively high attenuation of the longer wavelengths by the necessarily thicker than normal cell windows, has tended to limit the number of published applications. The safe containment of α-active materials invariably requires the use of a double container consisting perhaps of a primary P.V.C. covering on the sample plus a second container which can comprise the sample cell.[42] Heat sealed P.V.C. sachets can also be used for the primary containment of samples consisting of borax beads or pellets. $\beta\gamma$-active materials usually present less of a handling problem since the ingestion hazard is not so important. In this instance single containment will often suffice. The magnitude of the additional shielding required to prevent the stray radiation from entering the detector depends to a large extent upon the type and the magnitude of the activity of the sample.[42] Stray radiation entering the detector from directions other than that defined by the geometry of the spectrometer can be almost completely removed by means of suitably placed lead shielding. The real problem is to minimise scattered radiation which enters the detector along the same optical path as the X-radiation — the major source of this interference being due to radiation from the sample passing through the primary collimator and being scattered by the analysing crystal. Should this scatter consist of γ-radiation its effect can be almost completely removed by careful application of pulse height selection, since the energy of γ-radiation is usually an order of magnitude greater than the X-radiation being measured.

REFERENCES

* 1 BERTIN, E. P. and LONGOBUCCO, R. J., 1962, Norelco Reporter, 9, 31.
* 2 BERTIN, E. P. and LONGOBUCCO, R. J., 1965, Norelco Reporter, 12, 15.
3 ABRENS and TAYLOR, Spectrochemical Analysis, Addison-Wesley, Reading-Mass, 41.
4 LAMBERT, M. C., 1959, Norelco Reporter, 6, 37-51.
5 HANS, A., L'analyse spectrochimique par les rayonnements X, (Milan, 1961), Philips, Eindhoven, 29.
6 BERTIN and LONGOBUCCO, Advances in X-ray Analysis, Plenum, New York, 1961, 5.
7 MACDONALD, G. L., Proceedings of 4th M.E.L. Conference on X-ray Analysis, (Sheffield, 1964), Philips, Eindhoven, 11.
* 8 JENKINS, R. and HURLEY, P. W., Proceedings of XIIth International Spectroscopy Colloquium, Hilger and Watts, London, 1965, 444.
9 JOHNSON, W., Proceedings of 4th M.E.L. Conference on X-ray Analysis, (Sheffield, 1964), Philips, Eindhoven, 73.
10 CAMPBELL, W. J., LEON, M. and THATCHER, J. W., 1959, U.S. Bureau Mines Report No. 5497.
11 Encyclopedia of X-rays and γ-rays, New York, Reinhold, 1963.
* 12 ANDERMANN, G. and KEMP, J. W., 1958, Anal. Chem., 30, 1306-1309.
13 SARIAN, S. and WEART, H. W., 1963, Analyt. Chem., 35, 115.
* 14 CLAISSE, F., 1957, Norelco Reporter, 4, 3.
* 15 CLAISSE, F. and SAMSON, C., Advances in X-ray Analysis, Plenum, New York, 1961, 5, 335.
16 BLANQUET, P., L'analyse par Spectrographie et Diffraction de rayons X, (Madrid, 1962), Philips, Eindhoven.
17 A.R.L. Spectrographer's News Letter, 1954, 7, 1-3.
18 NORRISH, K. and HUTTON, J. T., 1964, C.I.S.R.O. Division of Soils, Adelaide, Divisional Report 3/64.
19 CARR-BRION, K. G., 1964, Analyst, 89, 556-557.
20 ANDERMAN and ALLEN, Advances in X-ray Analysis, Plenum, New York, 1960, 4, 414.
21 CAMPBELL and THATCHER, Advances in X-ray Analysis, Plenum, New York, 1958, 2, 313.
22 CULLEN, T. J., 1960, Analyt. Chem. 32, 516-7.
23 CULLEN, T. J., 1962, Analyt. Chem., 34, 862.

24 RINALDI, F., 1966, Scientific and Analytical Bulletin, No. 5, Philips, Eindhoven.
25 JENKINS, R., 1962, J. Inst. Pet., **48**, 246.
* 26 NATELSON, S. and BENDER, S. L., (1959), Microchemical Journal, **3**, 19.
27 JENKINS, R. and PRICE, N.B., (unpublished).
* 28 RHODIN, R. N., 1956, Norelco Reporter, **3**, 14-8.
29 JENKINS, R., 1963, Erdoel-Z. Bohr-u. Fordertech, **79**, 59-66.
30 BOOKER, G. R. and NORBURY, J., 1957, Brit. J. Appl. Phys., **8**, 109-13.
31 HIRT, R. C., DOUGHMAN, W. R. and GISCLARD, J. B., 1956, Analyt. Chem., **28**, 1649-51.
32 GRUBB, W. T. and ZEMANY, R. D., 1955, Nature, **176**, 221.
33 COLLIN, R. L., 1961, Analyt. Chem., **33**, 605-7.
* 34 VAN NIEKERK, J. N. and DE WET, J. F., 1960, Nature, **186**, 380-1.
35 VAN NIEKERK, J. N., DE WET, J. F. and WYBENGA, F. T., 1961, Analyt. Chem., **33**, 213-5.
36 MINNS, R. E., *Proceedings of Exeter Conference on Limitations of Detection in Spectro-chemical Analysis*, (July 1964), Hilger and Watts, London.
37 CAMPBELL, W. H., SPANO, E. F. and GREEN, T. E., 1966, Analyt. Chem., **38**, 987.
38 NATELSON, S. and SHEID, B., (1959), Clinical Chemistry, **5**, 519.
39 NATELSON, S. and SHEID, B., (1960), Clinical Chemistry, **6**, 299.
40 NATELSON, S. and SHEID, B., (1961), Clinical Chemistry, **7**, 115.
41 LIND, P. K. and MATHIES, J. C., (1960), Norelco Reporter, **7**, 127.
* 42 BREMNER, W. B., 1964, *Proceedings of 4th M.E.L. Conference*, (Sheffield, 1964), 65-71, (Philips, Einᵈ ioven).
43 DUBROVO, S. K., 1964, *Vitreous Lithium Silicates*, Nauka: Leningrad (English translation available).
44 DIJKSTERHUIS, P. R., 1969, Kalk, Zement und Gips (in press)

CHAPTER 9

TRACE ANALYSIS

9.1 General

Since X-ray fluorescence spectrometry is essentially a method which counts atoms, the question naturally arises as to what is the minimum number of atoms which are required in order to give a measurable signal above background. Analyses based on the measurement of a small number of atoms fit conveniently into two categories, the first where the number of analysed atoms is small in comparison with the total number of atoms making up the sample i.e. in the analysis of low concentrations and second, where the number of atoms is small because the total sample weight is small i.e. in the analysis of limited quantites of material. These two cases must be considered separately since, as will be seen later, they have little in common.

9.2 Analysis of low concentrations

In common with other methods of physico-chemical analysis the analysis of low concentrations by X-ray fluorescence spectrometry can be defined in terms of the ability to detect and measure a very weak response superimposed upon a background of almost equivalent strength. Compared with wet chemical analysis where the background corresponds to the so-called blank determination the true spectroscopic background corresponds to an absorption or emission at the same wavelength as that of the response being measured, but in the special case where the element or compound giving rise to an intensity change is absent. In spectrometry it is rarely possible to measure the true background intensity since by definition the weak signal corresponding to the low concentration occurs at the same wavelength. Moreover, the scattering power of a sample depends on its composition, it is thus very often not feasible to measure the background on the true line position in a sample of similar composition, but with the element in question missing. As a result it is invariably necessary to carefully select an empirical background wavelength which has the same response characteristics as that of the true background wavelength. This is particularly important in X-ray spectrometry where the presence of scattered X-ray target lines, weak harmonic reflections and matrix effects all demand special care in the selection of a suitable background wavelength.

If a linear plot is drawn of response plus background, against concentration of the element giving rise to that response, the sensitivity of the spectroscopic method will depend upon the slope of the curve and the level of the background. In general, the steeper the slope and/or the lower the background, the greater the sensitivity. Fig. 5.1 shows the response curve as it would occur in X-ray fluorescence analysis taking the simple case of a linear calibration curve — R_p is the counting rate at the peak position, R_b is the true background count rate, C is the concentration of the measured element and "m", the slope factor, is a measure of the sensitivity in counts per second per %. The reproducibility of R_p and R_b are of course dependent, within certain equipment limitations, upon the number of counts taken on peak and background positions. Since it has already been stated that trace analysis depends upon making measurements of R_p at count rates approaching R_b in the analysis of low concentrations by X-ray fluorescence analysis the three defining parameters are m the slope factor, R_b the background counting rate and $\sigma(R_p R_b)$ the precision of the peak and background count rates.

9.3 Theoretical considerations

Before proceeding further it may be as well to consider the theoretical treatment of the problem and to establish mathematical relationships between m, R_p and $\sigma(R_p R_b)$ in order to obtain working definitions of both lower limit of detection and a figure of merit that can be used to decide the point at which operating variables are optimal.

9.4 Statistical definition

The statistical definition of a lower limit of detection has been considered by many workers[1-4] and is normally defined as that concentration which gives a count rate equivalent to a background reading plus twice the standard deviation of the background, or $\bar{R}_b + 2\sigma_{R_b}$. This means that, if following such a reading, the presence of characteristic radiation is assumed, this assumption will be incorrect in 5 % of cases (95 % confidence limit). Where this uncertainty is considered to be too great a 3 σ interval, $\bar{R}_b + 3\sigma_{R_b}$ can be used. It is clear that a random error will be associated with any background measurements and if a certain confidence limit is placed on the magnitude of that error any data falling outside the calculable spread may be taken as being due to a different source from that of the background. If the background intensity is independent of the presence of the trace element a series of readings on a blank sample similar in composition will give the mean value of the background R_b and the standard deviation of an individual measurement. This is, however rarely possible in practice. In all other cases a separate measurement of the background has to be made which gives R_b with a standard deviation $\sqrt{\dfrac{R_b}{T_b}} = \sigma_{R_b}$. The detection limit is now increased by a factor of $\sqrt{\dfrac{R_p}{T} + \dfrac{R_b}{T}} : \sqrt{\dfrac{R_b}{T}}$

following from the rule for summation of total variance. This ratio can be taken as $\sqrt{2}$ since $R_p \approx R_b$. Thus in practice the limit of detection is taken as that concentration which gives a net count rate equivalent to 3 times $(2\sqrt{2})$ the standard deviation of the background count rate for the 95 % confidence limit.

9.5 Figure of merit (or quality function)

It follows from Chapter 5 and the foregoing, that

$$\sigma_{R_b} = \sqrt{\frac{R_b}{T_b}} \text{ or } 3\sigma_{R_b} = 3\sqrt{\frac{R_b}{T_b}} \text{ cps, which is by definition the}$$

minimum detectable net intensity.
Now the slope factor m expresses counts per second per percent element present in this range; thus the lower limit of detection will be given by.

$$l.l.d. = \frac{3}{m}\sqrt{\frac{R_b}{T_b}}\%, \text{ where } T_b \text{ is the counting time on the background}$$

i.e. one half of the total analysis time. For a fixed analysis time the figure of merit is taken as $m/\sqrt{R_b}$ and in order for the detection limit to be at a minimum the figure of merit should be as large as possible.

9.6 Generator stability

Owing to the fact that the production of X-rays is a random process any measurement which is made of the counting rate will be subject to errors other than those brought about by instrumental fluctuations. Thus inte-

gration of the signal over a fixed period of time is important not only from
the point of view of overcoming unpredictable instrumental drift but also
as a means of controlling within prescribed limits the predictable spread of
data due to the random distribution of the generated X-rays. Probably the
most important practical limitation in all quantitative X-ray spectrometry
arises of the point where instrumental drift becomes the limiting factor in
the precision of a measurement rather than the number of counts taken. It is
extremely important therefore that the exact point be established at which
the collection of further counts no longer improves the precision of the
measurement.

Considering first the measurement of one single intensity the standard
deviation of the measurement can be correlated with the number of counts
using the relationship $CV = \dfrac{100}{\sqrt{N}}$, where CV is the coefficient of variation.
A plot of log CV against log N shows a linear relationship as will be seen
in Fig. 5.3 and somewhere along the correlation line will be a point which
will represent the practical limitation of N. The actual position of this point
will depend a great deal on the type of spectrometer used and the method by
which it is employed. For example, it is possible by means of a sequential
ratio instrument that the only variables between measuring an unknown
sample and a standard are the time dependent variables such as E.H.T.
and of course time itself. The sequential-ratio system allows that systematic
errors such as angular setting, crystal setting and so on can be eliminated,
leaving errors due to fluctuations in the X-ray generator, the E.H.T. supply
to the detectors and the timing device. Of these the generator stability is
usually the limiting factor. A typical generator which is currently available
is one which gives a stabilisation of 0.03 % on both kV and mA provided
that the main supply fluctuations lie within certain limits. In general, the
intensity of any X-ray line is a linear function of tube current and approximately linear to a square power of kV, (cf Appendix 2b) so from this it can
be predicted that the overall variation in intensity from a certain X-ray line
would be of the order of 0.1 % for a single measurement or 0.14 % for a ratio
measurement. From Equation (5.7) it will be seen that this will correspond
to a number of counts $N = 10^6$.

9.7 Effect of long term drift

In the analysis of low concentrations the situation is rather more complicated because measurements have to be made at both peak and background

positions. In addition, due to the fact that counting rates are low, the times required for the measurements are relatively long and this introduces a further possible course of error arising from the effects due to long term drift. If a ratio measurement were made on a sample and standard in which 10^6 counts were accumulated from the standard in time t, and the counts accumulated from the sample in time t were measured, it would be found that on repeating the measurement many times although the count ratio data lie within the expected statistical spread, t might vary by as much as three times this amount. This effect stems from the fact that X-ray generators invariably have very high stabilisation on a purely short (of the order of minutes) term basis. When working at reasonably high count rates long term drift does not matter because the standard and sample sequence is being measured over a relatively short period of time where the short term drift is still the overiding influence. However, when working at low count rates, over long time integrations, long term drift itself can become the limiting factor. As a result it is necessary when working with trace concentrations to take care in the way in which time for the measurement of peak and background is divided. If for example analysis times are in excess of ten minutes, it is far better to split up the theoretical N_p and N_b (with due statistical adjustments) such that several alternate readings can be made sequentially rather than measuring the full N_p followed by the full N_b.

9.8 Variation of detection limit with atomic number

Fig. 9.1 shows the way in which the detection limit varies with atomic number for an average matrix. It will be appreciated that variation in the absorption characteristics of different matrices can cause wide variation in detection limits, since obviously, the lower limit of detection for a certain element will be far better in a low absorbing matrix than in a strongly absorbing matrix. This point will be dealt with in more detail later. First however, the general shape of the detection limit curve should be considered. The curve can best be considered in two distinct portions, the first between atomic numbers 12 and 56 and the second above atomic number 56. The second part is in fact almost an exact repeat of the first part, except that it is displaced upward by a factor of about 10. This is because whereas up to atomic number 56, K lines can be used to the best advantage, above this point L lines must be used which are considerably less intense than K lines owing to greater specimen absorbtion. Any remarks made about the detection limit curve up to atomic number 56 will apply similarly to the equivalent

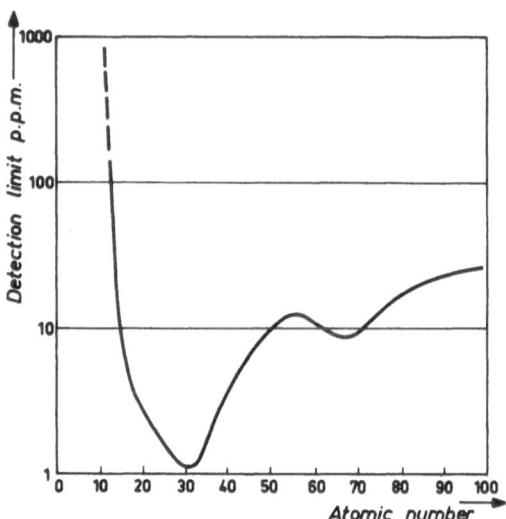

Fig. 9.1 Variation in detection limit with atomic number for an average matrix.

wavelength range above this point with the exception that sensitivities are down by about a factor of 10.

The rapid fall-off in sensitivity at the low atomic number end of the scale is due mainly to three factors. First the fluorescent yield values drop off rapidly to values of less than 0.1 below atomic number 20, second, much of the useful exciting X-radiation is attenuated by the window of the X-ray tube and the characteristic radiation from the sample is further absorbed by the detector window and third, the long wavelength of the characteristic radiation invariably requires the use of a rather inefficient analysing crystal. Unfortunately, owing to the limited range of analysing crystals available and to general wavelength resolution problems, there is frequently little choice in the selection of an optimum crystal for a specific trace analysis problem. However, one invariably has some choice in the selection of excitation conditions.

9.9 Choice of excitation conditions

9.9.1 CHOICE OF X-RAY TUBE

The increasing range of different anode materials and design of X-ray tubes allows a wide flexibility in the well equipped X-ray analytical laboratory.

Since much of this topic has already been covered in Chapter 1 a few examples of the variation in detection limit by use of different anode materials should serve to underline the major principles involved. Table 9.1 illustrates the improvement in the lower limit of detection for low atomic number elements which can be obtained by use of the chrome anode tube compared with the tungsten anode tube. The first example is of silicon in steel and it will be seen that whereas the detection limit for a tungsten anode tube is 0.03 %, a three fold decrease can be obtained by use of the chromium anode tube. The second example is that of sodium in glass and here the improvement in detection limit is also of the order of 3 to 4 times.

TABLE 9.1

Detection limits with Cr and W anode tubes (40 seconds analysis time)

1. Si in Steel

        0.3% Si gave the following data:

a) W anode tube:

    $R_p = 100$ c/s $m = 160$ c/s per %
    $R_b = 52$ c/s   detection limit $= 0.03$ %

b) Cr anode tube:

    $R_p = 86$ c/s $m = 247$ c/s per %
    $R_b = 12$ c/s   detection limit $= 0.009$ %

2. Na in glass

        10.8% Na in glass gave the following data:

a) W anode tube:

    $R_p = 14.5$ c/s $m = 0.93$ c/s per %
    $R_b = 4.5$ c/s   detection limit $= 1.53$ %

b) Cr anode tube:

    $R_p = 63$ c/s $m = 4.9$ c/s per %
    $R_b = 10$ c/s   detection limit $= 0.43$ %

9.9.2 CHOICE OF TUBE CURRENT AND POTENTIAL

The choice of tube current is not critical since both the peak and background count rates increase more or less linearly with increase of current. The choice of tube potential is, however, very much more important since this will have a different effect on the peak and background counting rates. It will be seen from Equation (5.15) that the figure of merit for optimum

TABLE 9.2

Selection of optimum kV

kV (V_0)	Al ($Z = 13$)	Cr ($Z = 24$)	Zn ($Z = 30$)	Mo($Z = 42$)	In ($Z = 49$)
50	100	99	96	67	43.5
60	98	100	100	80	61
70	92	97	100	87	75
80	88	94	98	95	84
90	—	89	95	97	93
100	83	—	—	100	100
V_0 (kV)	1.6	6.0	9.7	19.1	28.1

Mo target tube

Figure of Merit $\sqrt{R_p} - \sqrt{R_b}$ (cf section 5.8) Normalised to best $= 100$

intensity if $\sqrt{R_p} - \sqrt{R_b}$ and Table 9.2 compares the figures of merit obtained for a range of critical excitiation potentials (V_c) and tube potentials (V_0). All data have been normalised in such a way that the best figure is equal to 100. The data show that the optimum value of V_0 increases steadily from 50 kV for aluminium ($Z = 13$) to 100 kV for molybdenum ($Z = 42$) and it is apparent from this that for wavelengths shorter than about 0.8 Å the optimum tube potential setting for 100 kV maximum constant potential generator is at maximum. However, this is purely from the point of view of the slope factor. In order to study the effect of detection limits, the effect of

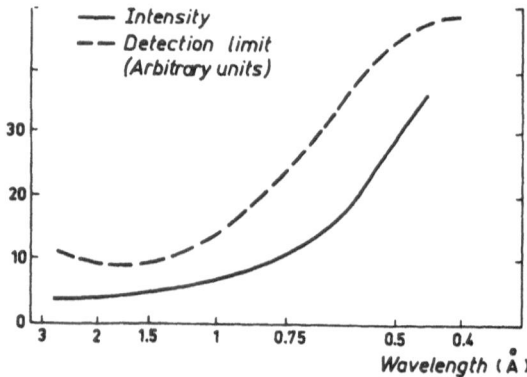

Fig. 9.2 Comparison of scattered background and detection limit. The full curve illustrates the shape of the scattered background obtained using a pure aluminium sample and a tungsten target X-ray tube operated at 48 kV. The dotted curve represents the average detection limits obtainable over the same range.

tube potential on R_b should also be considered. Fig. 9.2 shows a section of the detection limit curve again along with a typical background plot for the same wavelength range. It will be seen that the shapes of the curve are almost identical.

9.10 Effect of background

The background consists mainly of coherently and incoherently scattered primary radiation along with a certain amount of characteristic radiation from the sample itself. The scattered primary radiation consists of both continuous and characteristic and it is the continuous which gives the background its distinctive humped appearance. To the high atomic number end of the spectrum the background is mainly first order·scatter but at low energies higher orders provide the main contribution. It is for this reason that energy dispersion, in the form of pulse height selection, which is so successful for removing background and spectral interferences at low energies, is of little use for the higher energy radiations. Due to the inherently poor resolution of detectors separation of first order characteristic radiations is impossible for elements of less than 5 atomic numbers apart in this part of the range. The background falls off rapidly to zero at the wavelength corresponding to the potential on the tube after peaking at about twice this value. Thus although the background curve always has approximately the same shape it may be displaced to a higher or lower energy depending upon the tube potential. The overall intensity of the background is a function of the scattering power of the sample which in turn depends upon the average atomic number of the matrix. Many methods have been applied to aid in the reduction of low order high intensity background, of which, probably only two are worth mentioning.

9.11 Remqval of background by polarisation [5)

An inherent property of the X-ray fluorescence system stems from the fact that background radiation is plane polarised after reflection by the sample. The degree of polarisation depends upon the angle between the primary and secondary directions; for most equipment this angle is close to 90°, so the degree of polarisation is high. Characteristic radiation excited in the sample is however, not plane polarised. Both background and characteristic radiation are plane polarised after reflection by the crystal. The scattered background radiation can pass the crystal if the planes of polarisation of the sample and the crystal are parallel. Unfortunately, this situation exists in most spectrometers.

However, the background intensity can be much reduced if the polarisation plane of the crystal is perpendicular to that of the sample. The degree of polarisation depends again on the scattering angle of crystals 2θ, and it can be shown theoretically that this degree of polarization will be at a maximum when $2\theta = 90°$.

Measurements have recently been made to investigate the feasibility of this approach using a conventional spectrometer in which the relative positions of the X-ray tube and spectrometer components were changed by rotating the X-ray tube and sample holder through 90° along the axis of the primary collimator. The *plot* of the relative background intensities using normal optics and polarising optics against 2θ shown in Fig. 9.3 illustrates that maximum background suppression is indeed achieved at around $2\theta = 90°$

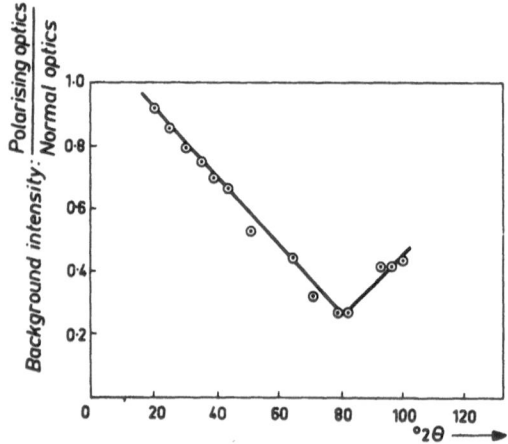

Fig. 9.3 Effect of polarising optics on background. The plot shows the correlation between relative background intensities, using polarising and non-polarising optics, against $°2\,\theta$. It can be predicted that maximum background suppression should occur at $90°2\theta$.

Background ratios are found to be better by a factor of about two around the 90° (2θ) position of samples with a high scattering power.

Unfortunately, the practicability of employing this kind of optical geometry is limited owing to the need for the sample cell to remain vertical and the specimen to be rotated during exposure. In addition, a convenient system of changing specimens is necessary. There are, however, certain instances where the system does offer potential advantages as, for example, in applications of "on-stream" analysis where special sample cells have to be designed.

9.12 Use of filters

The use of filters has been applied with particular success in the removal of background interference due to scattered tube lines. The scattered radiation from the sample contains not only continuous characteristic radiation from the tube anode but in addition, contains small contributions from tube impurities[6] including Fe, Cr, Cu and Ni. Complete spectral purity is difficult and costly to achieve and although spectral contamination has decreased steadily over the years, significant contributions from these four elements still occur and can be particularly troublesome when trying to detect low concentrations of these elements in low atomic number matrices. For example, the contaminant lines can be equivalent to as much as 5 to 10 parts per million of the elements in a matrix of pure aluminium. This situation poses two problems in trace analysis, first that the true background can be as much as twice the predicted value and second, that it is difficult to assess exactly what the true background contribution really is.

Data from a typical X-ray tube shows that the detection limit for iron in aluminium is 1.5 ppm but without the iron contribution from the tube this value would be 0.4 ppm. These tube lines can however by removed by making use of suitable filters between the X-ray tube and sample and the method has been applied successfully where suitable filter thicknesses are available.[7-8] For example, in the analysis of chromium in low atomic number matrices using a chrome anode X-ray tube the detection limit is of the order of 2.8 ppm. By making use of a 0.005 inch thick titanium filter between the X-ray tube and sample the tube contribution is completely removed and the detection limit is lowered by a factor of 3. One difficulty however is that frequently suitable filter materials are not available and, even if they were, problems are frequently encountered in producing a filter of the correct thickness.

9.13 Effect of the matrix

Finally the effect of the sample matrix itself on the detection limit should be considered. As was stated earlier the detection limit for a certain element

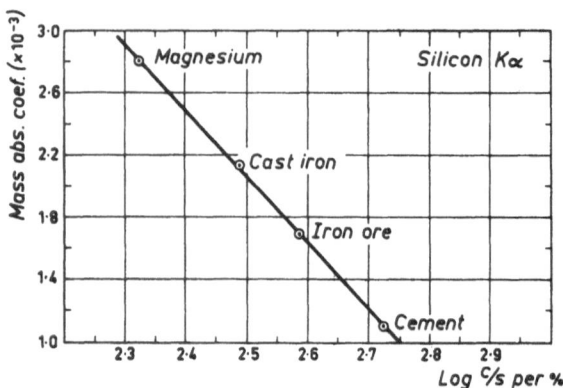

Fig. 9.4 Effect of matrix absorption on sensitivity. A plot of mass absorption coefficient μ against log. c/s per % for silicon Kα radiation shows linear correlation over a wide range of materials.

will be dependent upon the matrix absorption. For example, taking tin as a typical case, Table 9.3 lists data obtained from a range of tin containing matrices. It will be seen that the figure of merit described earlier $\dfrac{m}{\sqrt{R_b}}$ is of the same order for both the relatively high atomic number matrices of mild steel and 60/40 brass. The detection limit in each case is 13 parts per million. For the low atomic number matrices, however, m increases at a greater rate than R_b and the figure of merit increases by a factor of 3 giving a detection limit of 4 parts per million. It would be very useful to be able to predict the detection limits for specific elements in different matrices and this is certainly possible in the case of low atomic number elements where R_b when measured with pulse height selection is reasonably constant and the only effect of the matrix is to change m.[9-10] For example, in the case of silicon R_b varies between 9 and 14 c/s for a wide range of matrices including steels, copper based, aluminium based materials and slags. From the standard absorption equation $I = I_0 \exp(-\mu\rho x)$, m would be expected to give a linear log relationship with μ. Fig. 9.4 shows a plot of log m against μ for a range of matrices and demonstrates that in general, extremely good correlation is shown. Thus knowing for example the detection limit of silicon in cast iron, it is possible to predict what the detection limit of silicon will be in magnesium merely by multiplying the detection limit in iron by the ratio of the matrix mass absorption coefficient values. In the case of iron the detection limit is 56 ppm thus in magnesium it will be $56 \times \dfrac{2150}{2800} = 43$ ppm. In fact the measured limit is 45 ppm. Unfortunately it is impossible to apply this method so successfully for the higher atomic number elements since in their characteristic line position the background is more variable.

TABLE 9.3

Effect of matrix absorption on detection limits for tin

Matrix	m (c/s per %)	R_b (c/s)	$m/\sqrt{R_b}$	l.l.d. (%)
Mild Steel	1940	72	229	0.0013
Brass (60/40)	2660	140	247	0.0013
Aluminium	21100	720	788	0.0004
Magnesium	27300	1120	816	0.0004

9.14 Analysis of limited quantities of material

The inherent limitations in the technique of X-ray spectroscopy which have been discussed in the previous section, often fix the lower limit of detection for a certain element in a specific matrix, at a level which is unacceptably high for the analytical chemist. When this situation occurs some form of concentration technique has to be employed in order to increase the percentage of required atoms to such a level that they fall within sensitivity range of the X-ray method. Many of these concentration methods have already been discussed in the preceeding chapter on sample handling. In general, where concentration techniques have been employed the analysed sample can no longer be treated as a problem in the determination of a low concentration of an element in a large sample, but has to be considered in terms of the determination of a high concentration of an element in a limited total quantity of sample. Other types of sample can also be considered as fitting into this general classification and undoubtedly the most important of these fall under the broad heading of thin films. Other analyses which should be treated as the estimation of elements in limited quantities of material, include the examination of small selected area of large samples, as in the application of macroprobe X-ray[11] and electron probe microanalysis; since this latter technique falls outside the terms of reference of this book the reader is recommended to consult other works dealing exclusively with this topic.[12]

9.15 Theoretical considerations

Section 7.4 discussed the theoretical basis for the thin film method, which in simple terms can be expressed as

$$I_m^i = W^i \times I_i^i \qquad (9.1)$$

which simply states that the intensity of the ith element in matrix m is equal to the product of the weight fraction of element i and the intensity obtained from the pure element. In other words, when matrix effects are negligible the intensity of an X-ray line is directly proportional to the number of atoms giving rise to that line. This situation is never achieved in massive samples where absorption effects on both primary and characteristic radiation are inevitable. However, when the sample thickness approaches the same order of magnitude as the path length of the measured radiation, the relationship given in Equation (9.1) is upheld. Thus it is apparent that there is a considerable incentive for use of thin samples purely from the point of view of avoiding matrix effects, and in point of fact, many quantitative methods of analysis have been based on this precept.[13-18] In the analysis of limited quantities of material it is frequently possible to prepare the sample as a thin film and indeed this is often the quickest and most reproducible method of sample preparation available. Thin films may be prepared in a variety of ways, for example, by evaporating aqueous solutions onto mylar film,[19-20] filter paper discs,[21] or poly-vinyl alcohol pellets, or perhaps by spreading powder samples onto scotch tape which can be fitted directly into the sample cup.[17] Quantitative measurements can then be made either by direct comparison with suitable standards or where these are not available, by use of the "spiking" procedure outlined in Chapter 7.

9.16 Ultimate requirements in sample size

The minimum sample size that is required to give a detectable signal above background usually falls within the range 0.01-1 microgram.[22] The actual value will depend to a certain extent on the substrate onto which the sample has been deposited because the background intensity is itself dependent upon the scattering power of the sample. Inasmuch as the sample thickness is at a minimum the majority of the scattering will arise from the substrate and the methods of background removal which have already been discussed in the previous section will apply equally well to the analysis of small samples.

9.17 Handling of small samples

In the case where a few milligrams of a sample are submitted for analysis the question arises as to the best method of presenting the sample to the spectrometer. From the foregoing it is obviously best to present the sample as a thin film if at all possible and for powders this can be easily achieved by spreading over scotch tape. A better method, however, is to dissolve the weighed sample in the minimum quantity of a suitable solvent and then to evaporate the solution onto a filter paper disc. A known weight of internal standard can also be deposited on the filter paper by spotting a known volume of a standard aqueous solution by means of a micropipette. Where the sample consists of a small pellet of material which for some reason cannot be destroyed, great care must be taken in the way in which the sample is presented to the spectrometer. The intensity distribution of the primary radiation across the area of the sample cell window is extremely variable and, for reasons of geometry, both the characteristic line intensity as well as goniometer angle setting will be dependent upon the actual position of the sample in the sample cup, see Fig. 8.6. Various methods for clamping minute samples in a sample cup in a reproducible manner have been described [23] and provided that the analysing crystal is not twinned this method of sample handling proves quite adequate. It should be pointed out however, that the intensities which are obtainable with a fixed quantity of sample taken as a small lump are invariably lower by several orders of magnitude compared to the same sample taken as a thin film. This effect will obviously become more pronounced as the path length of the characteristic radiation in the sample becomes shorter, since for a lump of material only small fraction of the total sample volume can contribute to the measured intensity.

REFERENCES

1 GARTON, F. W. G., United Kingdom Atomic Energy Authority report A.E.R.E. R-4483.
* 2 KAISER, H. and SPECKER, H., 1956, Z. analyt. Chem., **149**, 46.
3 KAISER, H., 1966, Z. analyt. Chem., **216**, 80.
* 4 ZEMANY, P. D., PFEIFFER, H. G. and LIEBHAFSKY, H. A., 1959, Analyt. Chem., **31**, 1176.
5 CHAMPION, K. P. and WHITTEM, R. N., 1963, Nature, **199**, 1082.
6 LADELL, J. and PARRISH, W., 1959, Philips Research Reports, **14**, 401.
7 CHAMPION, K. P. and WHITTEM, R. N., Report AAEC/TM289, Sydney, April 1965.
* 8 SALMON, *Advances in X-ray Analysis*, Plenum, New York, 1963, **6**, 301.
* 9 JENKINS, R., *Proceedings of Exeter Conference on Limitations of Detection in Spectrochemical Analysis*, Hilger and Watts, London, 1964.
10 MÜLLER, R., 1964, Spectrochim. Acta, **20**, 143-151.
11 LUKE, C. L., 1964, Analyt. Chem., **36**, 318.
* 12 BIRKS, *Electron Probe Micro-analysis*, Interscience, New York, 1963.

13 STONE, R. R. and POTTS, K. T., 1963, Norelco Reporter, **10**, 94.
14 SCHREIBER, T. P., OTTOLINI, A. C. and JOHNSON, J. L., 1963, Appl. Spectr., **17**, 17.
*   15 SALMON, *Advances in X-ray Analysis*, Plenum, New York, 1962, **5**, 389.
16 RHODIN, T. N., 1955, Analyt. Chem., **27**, 1857.
17 ADDINK, N. W. H., 1959, Rev. Universelle des Mines, **15**, 530.
18 NORRISH, K. and SWEATMAN, T. R., 1962, Divisional Report 11/61, C.S.I.R.O. Division
     of Soils, Adelaide.
19 PFEIFFER, H. G. and ZEMANY, P. D., 1954, Nature, **174**, 397.
20 GUNN, E. L., 1961, Analyt. Chem., **33**, 921.
21 JOHNSON, J. L. and NAGEL, B. E., 1963, Microchemica Acta, **3**, 525.
22 CAMPBELL, W. J. and THATCHER, J. W., 1962, U.S. Bur. Mines Rept. Invest 5966.
23 MACDONALD, G. L., *Proceedings of 4th M.E.L. Conference on X-ray Analysis*, (Sheffield,
     1964) Philips, Eindhoven.

Table of Mass-absorption Coefficients

Z	0.10	0.15	0.20	0.25	0.30	0.35	0.40	0.45	0.50	0.55	0.60	0.65	0.70	0.75	0.80	0.85	0.90	0.95	1.00	1.0
1 H	0.28	0.30	0.33	0.34	0.35	0.36	0.36	0.36	0.37	0.37	0.37	0.38	0.38	0.38	0.39	0.39	0.39	0.39	0.39	
2 He	0.14	0.16	0.17	0.17	0.18	0.18	0.18	0.19	0.19	0.19	0.20	0.20	0.20	0.21	0.22	0.22	0.23	0.24	0.24	
3 Li	0.12	0.14	0.14	0.15	0.16	0.16	0.17	0.17	0.18	0.19	0.19	0.20	0.22	0.23	0.24	0.27	0.29	0.30	0.32	
4 Be	0.13	0.14	0.15	0.16	0.17	0.17	0.18	0.19	0.21	0.22	0.24	0.25	0.29	0.32	0.36	0.37	0.42	0.48	0.53	
5 B	0.13	0.15	0.16	0.17	0.18	0.19	0.21	0.23	0.25	0.28	0.30	0.34	0.40	0.43	0.49	0.57	0.63	0.70	0.78	
6 C	0.14	0.16	0.17	0.19	0.21	0.22	0.26	0.28	0.34	0.38	0.45	0.53	0.62	0.72	0.81	0.96	1.15	1.30	1.40	
7 N	0.14	0.16	0.18	0.20	0.23	0.27	0.31	0.37	0.44	0.55	0.63	0.80	0.92	1.05	1.25	1.45	1.70	1.95	2.20	
8 O	0.14	0.17	0.19	0.22	0.26	0.33	0.38	0.49	0.58	0.72	0.87	1.10	1.30	1.50	1.80	2.10	2.45	2.80	3.30	
9 F	0.14	0.17	0.19	0.23	0.29	0.36	0.45	0.55	0.72	0.88	1.10	1.30	1.60	1.95	2.40	2.80	3.25	3.75	4.50	
10 Ne	0.15	0.18	0.22	0.27	0.36	0.43	0.59	0.72	0.98	1.20	1.55	1.90	2.30	2.80	3.45	3.95	4.70	5.50	6.50	
11 Na	0.15	0.18	0.23	0.29	0.40	0.52	0.71	0.93	1.20	1.45	1.95	2.45	3.00	3.60	4.40	5.25	6.15	7.20	8.35	
12 Mg	0.15	0.19	0.25	0.35	0.48	0.66	0.91	1.20	1.50	2.05	2.60	3.30	4.10	4.85	5.90	7.10	8.40	9.80	11.0	
13 Al	0.15	0.20	0.27	0.39	0.56	0.78	1.10	1.45	1.95	2.55	3.25	4.10	5.05	6.10	7.30	8.85	10.5	12.5	14.0	
14 Si	0.16	0.22	0.31	0.45	0.67	0.96	1.35	1.80	2.40	3.15	4.05	5.20	6.40	7.75	9.20	10.0	13.5	15.5	17.5	
15 P	0.16	0.23	0.34	0.51	0.77	1.15	1.55	2.20	2.90	3.85	4.85	6.15	7.70	9.35	11.0	13.5		18.0	21.0	
16 S	0.17	0.25	0.39	0.61	0.93	1.40	1.95	2.70	3.60	4.65	6.10	7.40	9.10	11.5	14.0	16.0	18.5	22.5	26.5	
17 Cl	0.17	0.26	0.42	0.68	1.05	1.65	2.25	3.25	4.20	5.50	7.05	8.85	11.0	13.5	16.0	19.0	23.0	26.0	30.5	
18 Ar	0.17	0.27	0.45	0.73	1.15	1.75	2.50	3.50	4.50	6.10	7.85	9.60	11.5	14.5	18.0	21.5	25.5	29.5	34.0	
19 K	0.19	0.31	0.54	0.91	1.45	2.20	3.15	4.40	5.95	7.80	10.0	12.5	15.0	18.5	23.0	27.5	32.5	37.0	43.0	
20 Ca	0.20	0.35	0.63	1.05	1.70	2.60	3.80	5.30	7.10	9.30	12.0	15.0	18.5	22.0	27.5	32.5	38.5	45.0	52	
21 Sc	0.20	0.37	0.67	1.10	1.80	2.65	3.85	5.40	7.30	9.60	12.5	15.5	19.5	23.5	28.0	32.0	39.0	46.0	53	
22 Ti	0.20	0.39	0.73	1.25	2.05	3.25	4.60	6.40	8.70	11.5	14.5	18.5	22.0	27.0	33.5	40.0	47.0	54	62	
23 V	0.22	0.42	0.80	1.30	2.15	3.30	4.85	6.80	9.10	12.0	15.5	19.0	23.5	28.5	34.5	41.5	48.5	57	65	
24 Cr	0.23	0.47	0.91	1.60	2.65	4.05	6.00	8.35	11.5	15.0	19.0	23.0	29.0	36.0	43.0	49.0	58	67	80	
25 Mn	0.24	0.50	1.00	1.70	2.70	4.10	6.00	8.40	11.5	15.0	19.0	24.0	29.5	36.5	43.0	52	61	71	83	
26 Fe	0.26	0.56	1.15	2.05	3.35	5.10	7.65	10.5	14.5	19.0	24.0	27.5	38.5	45.5	54	64	75	85	99	
27 Co	0.27	0.59	1.25	2.10	3.50	5.40	7.90	11.0	15.0	20.0	25.5	32.0	40.0	49.0	58	69	82	95	110	
28 Ni	0.30	0.69	1.40	2.60	4.30	6.70	9.70	14.0	18.0	23.5	30.5	37.5	48.0	58	67	74	87	105	122	
29 Cu	0.30	0.72	1.50	2.75	4.55	7.05	10.5	14.5	19.5	24.5	32.0	39.5	51	61	70	85	99	116	127	
30 Zn	0.32	0.79	1.65	3.05	5.05	7.75	11.5	16.0	21.5	28.5	35.5	44.5	53	65	77	97	112	122	138	
31 Ga	0.35	0.84	1.75	3.30	5.45	8.30	12.0	17.0	22.5	30.5	37.5	47.0	55	68	82	100	119	127	143	
32 Ge	0.35	0.89	1.90	3.50	5.85	9.00	13.0	19.0	24.5	31.5	40.5	48.5	60	72	87	107	126	133	152	
33 As	0.39	0.98	2.05	3.85	6.35	9.70	14.0	20.5	26.5	34.5	43.5	55	64	76	91	113	133	141	160	
34 Se	0.39	1.00	2.15	4.05	6.80	10.5	15.0	21.5	28.0	36.5	46.0	56	69	83	97	113	142	151	26	
35 Br	0.42	1.10	2.45	4.50	7.50	11.5	16.5	23.0	31.0	39.5	50	62	76	90	105	127	150	235	27	
36 Kr	0.44	1.20	2.60	4.80	7.95	12.5	17.5	24.5	32.5	42.0	53	65	80	95	109	134	22	26	30	
37 Rb	0.47	1.30	2.80	5.20	8.65	13.5	19.0	26.0	35.0	45.0	57	70	86	101	116	2050	2450	29	33	
38 Sr	0.50	1.40	3.05	5.60	9.35	14.0	20.5	28.5	37.5	48.0	60	75	92	109	19	2250	2650	31	36	
39 Y	0.54	1.55	3.25	6.05	9.95	15.0	22.0	30.0	40.0	50	65	79	99	16	2050	2450	29	34	40	
40 Zr	0.57	1.65	3.50	6.50	10.5	16.0	23.5	33.0	42.5	53	68	84	15	19	2250	27	32	38	43	
41 Nb	0.61	1.75	3.75	6.96	11.5	17.0	25.0	33.5	45.0	57	73	89	16	20	24	2850	34	40	48	
42 Mo	0.65	1.85	4.05	7.45	12.5	18.5	27.0	35.5	48.0	60	75	15	19	22	26	31	3650	43	51	
43 Tc	0.67	2.00	4.25	8.00	13.0	20.0	28.5	38.0	51	63	10.5	15	19	2350	28	33	4050	46	56	
44 Ru	0.69	2.00	4.50	8.50	14.0	20.5	29.5	40.0	53	66	13	16.50	20	2550	30	36	44	49	59	
45 Rh	0.73	2.15	4.80	9.00	14.5	21.5	31.5	41.5	56	12	15	19	23	28	3350	3950	4650	54	62	
46 Pd	0.80	2.30	5.15	9.45	15.5	23.0	33.0	44.0	57	13	16.50	2050	25	27	31	36	4250	50	58	66
47 Ag	0.86	2.35	5.50	10.0	16.5	24.5	35.0	47.0	10.50	13.50	17.50	22	27	33	39	46	54	63	73	
48 Cd	0.86	2.45	5.55	11.0	17.5	25.0	35.5	49.0	11.00	14.50	18.50	2350	29	35	4150	49	5750	67	77	
49 In	0.89	2.60	5.80	10.5	18.0	26.0	37.0	9.00	12.00	15.50	20	25	3050	3750	44.50	4450	61	71	82	
50 Sn	0.97	2.85	6.30	11.5	18.5	27.0	38.5	9.50	13.00	16.50	21	2650	3250	40	47	55	65	76	86	
51 Sb	0.99	2.95	6.35	12.0	19.0	28.5	40.0	10.00	13.50	17.50	22.50	28	34.50	43	50	59	6950	81	93	
52 Te	1.05	3.10	6.75	12.5	20.0	29.0	7.5	10.50	14.00	18	23	2850	35	44	51	6050	71	83	95	
53 I	1.13	3.35	7.30	13.0	21.0	31.0	8.10	11.50	15.50	19.50	25	31	38	4650	55	6450	76	88	98	
54 Xe	1.15	3.45	7.55	13.5	22	32.5	12	16	20.5	26.5	32.5	4.05	49	58	69	80	93	112		
55 Cs	1.20	3.55	7.60	14.0	23	6.20	9	12.5	17	22	28	34.5	43	52	61	73	85	99	116	
56 Ba	1.30	3.80	7.75	14.5	23.5	6.50	9.40	13.5	18	23	29.5	37	45	56	65	77	90	104	119	
57 La	1.30	3.80	8.00	14.5	24	6.90	10	14	19	24.5	31	38.5	48	59	69	81	95	111	126	
58 Ce	1.30	3.80	8.20	14.5	24.5	7.40	11	15	20	26	33	41.5	51	61	73	86	100	117	135	
59 Pr	1.35	4.00	8.65	15.5	5.10	7.80	11.5	16	21	27.5	35.5	43.5	54	66	78	91	105	124	143	
60 Nd	1.40	4.15	8.90	15.5	5.40	8.30	12	16.5	22.5	29	36.5	47.5	56	68	81	94	110	130	149	
61 Pm	1.50	4.25	9.25	16	5.60	8.70	12.5	17.5	23.5	30	38	48.5	60	71	84	100	115	135	156	
62 Sm	1.56	4.50	9.55	16.5	5.70	8.80	12.5	18	23.5	31	39	48.5	60	73	86	105	120	137	158	
63 Eu	1.65	4.70	9.85	3.70	6.20	9.50	14	19	25.5	33	42	53	65	78	92	110	125	147	169	
64 Gd	1.70	4.85	10	3.90	6.50	9.80	14	20	26	34	43.5	54	66	81	95	115	130	153	176	
65 Tb	1.75	5.10	10.5	4.10	6.70	10	14.5	20.5	27	35.5	45	56	69	84	100	117	135	158	180	
66 Dy	1.85	5.25	11	4.30	7.00	10.5	15	22	28	36.5	47	58	71	86	102	122	140	164	187	
67 Ho	1.95	5.45	11	4.40	7.20	11	16	22	29.5	38.5	48.5	61	75	92	107	127	148	173	198	
68 Er	2.00	5.65	11.5	4.50	7.50	11.5	16.5	23	31	40.5	51	63	78	96	113	132	157	180	202	
69 Tm	2.10	5.80	12	4.70	7.80	12	17.5	24.5	32.5	42.5	54	67	83	101	118	141	165	192	211	
70 Yb	2.15	6.10	12	5.00	8.20	12.5	18	25	33.5	43.5	55	69	86	104	123	145	168	197	221	
71 Lu	2.25	6.30	3.00	5.40	8.70	13	19	26.5	35	46	59	73	89	109	128	153	178	204	235	
72 Hf	2.30	6.50	3.30	5.70	9.30	13.5	20	27.5	37	47.5	61	75	93	115	135	160	186	210	245	
73 Ta	2.35	6.70	3.40	6	9.50	14	20.5	28.5	38	49.5	63	79	97	121	140	166	195	220	250	
74 W	2.45	6.90	3.50	6.20	9.90	14.5	21.5	30	40	52	66	83	102	127	148	174	201	230	260	
75 Re	2.55	7.25	3.60	6.30	10	15	22.5	31.5	42	54	70	87	107	133	155	183	210	245		
76 Os	2.65	7.45	3.65	6.50	10	15.5	22.5	31.5	42	55	71	91	113	138	162	192	225			
77 Ir	2.70	7.65	3.75	6.70	10.5	16	23	33	44	57	72	91	113	138	162	192	225			
78 Pt	2.80	7.75	4.25	7.30	11.5	17	24.5	34	47	58	73	92	115	140	165	195	165	146	165	
79 Au	2.90	8.00	4.40	7.45	11.5	17.5	25.5	35	48.5	61	77	98	125	150	170	200		156	178	
80 Hg	3.00	2.35	4.55	7.90	12	18	26.5	37	49.5	65	82	103	125	150	185				180	
81 Tl	3.10	2.45	4.80	8.20	12.5	19	27.5	38.5	51	67	85	107	130	155	190				81	
82 Pb	3.15	2.50	4.90	8.25	13	19.5	28	39	53	67	87	108	135	160		125	140		75	89
83 Bi	3.20	2.55	5.10	8.60	13.5	20.5	30	41.5	56	73	92	117	140	165				79	91	10
90 Th	3.25	2.80	5.60	9.40	14.5	22	32	44	59	76	96	76	90	105		75	85	100	116	136
92 U	3.45	3.25	6.40	11	17.5	27	39.5	55	74	96			80			95	110	127	148	
	0.10	0.15	0.20	0.25	0.30	0.35	0.40	0.45	0.50	0.55	0.60	0.65	0.70	0.75	0.80	0.85	0.90	0.95	1.00	

	1.15	1.20	1.25	1.30	1.35	1.40	1.50	1.60	1.66	1.79	1.93	2.10	2.20	2.28	2.37	2.46	2.54	2.68	2.76	2.89	3.05	3.15
40	0.40	0.41	0.41	0.42	0.42	0.43	0.43	0.44	0.45	0.46	0.48	0.50	0.53	0.54	0.55	0.56	0.57	0.61	0.66	0.70	0.74	0.7
26	0.27	0.28	0.30	0.31	0.32	0.33	0.37	0.39	0.41	0.46	0.54	0.64	0.71	0.78	0.84	0.91	1.00	1.15	1.25	1.40	1.65	1.8
36	0.38	0.43	0.46	0.50	0.54	0.58	0.67	0.75	0.81	1.00	1.25	1.50	1.75	1.90	2.10	2.30	2.55	3.00	3.20	3.70	4.45	5.0
63	0.70	0.80	0.84	0.90	1.00	1.15	1.40	1.65	1.80	2.15	2.70	3.30	3.80	4.25	4.70	5.35	5.95	7.20	7.80	9.15	10.5	11
98	1.10	1.20	1.40	1.60	1.80	2.05	2.20	2.90	3.20	3.90	4.60	5.70	6.45	7.30	8.00	8.90	9.80	12	12.5	14.5	16.5	19
80	2.00	2.30	2.55	2.80	3.10	3.40	4.25	4.90	5.40	6.70	9.00	11.50	13.5	14.5	17.5	18	31	24	26	29	33	37
80	3.20	3.65	4.10	4.60	5.10	5.60	6.95	8.25	9.15	11.5	14.5	18.5	22.0	25.0	28.0	31	34	39	42	47	55	60
25	4.85	5.55	6.20	6.95	7.75	8.55	10.5	13.0	14.0	17.5	21.5	28.0	32.0	36.0	43	44.5	50	58	63	73	84	91
80	6.65	7.60	8.55	9.55	10.5	12.0	14.5	17.5	22.0	27.0	31.0	39.0	45.0	51.0	56	61	67	78	84	99	118	130
65	9.70	11.0	12.0	13.5	15.0	17.0	21.0	27.0	30.0	37.0	45.0	56	64.0	72	79	88	97	113	122	138	165	179
0	12.5	14.0	15.5	17.5	19.5	22.0	27.0	33.0	36.0	46.0	61	72	83.0	92	106	110	128	146	163	183	211	227
0	17.0	19.0	21.5	24.0	26.5	30.0	37.0	43.0	48.0	60	77	95	110	122	137	152	163	193	212	241	280	304
5	21.0	23.5	26.5	30.0	33.5	37.5	45.0	54	58	73	94	117	136	153	170	189	206	241	263	295	344	373
5	26.5	29.5	33.5	37.5	42.0	47.0	57	69	76	94	116	146	172	193	217	241	260	302	328	369	426	464
0	31.0	36.0	39.5	44.0	49.5	55	68	80	91	113	141	177	200	227	260	284	311	354	389	432	503	548
0	39.0	43.0	49.5	55	61	68	85	100	112	139	173	217	246	271	316	333	364	419	453	509	593	648
5	45.5	52	58	64	72	80	98	115	126	158	198	245	282	314	348	376	420	472	512	568	670	739
0	51	58	65	72	80	89	108	130	141	174	235	270	308	343	380	421	455	523	557	639	737	800
	65	73	82	92	103	114	136	167	179	218	269	330	380	423	514	552	632	689	770	891	968	
	78	86	99	110	123	137	161	199	210	257	305	400	440	483	532	583	631	719	780	875	1009	118
	79	89	101	113	125	140	168	201	222	273	338	421	483	540	595	668	732	850	98	111	127	138
	91	104	114	127	142	167	190	220	247	304	377	475	510	565	623	685	93	105	114	127	147	160
	99	112	125	140	155	172	210	250	275	339	422	530	612	77	88	96	106	120	129	146	167	181
	115	131	145	162	180	195	236	274	316	369	445	71	81	90	101	113	122	139	153	170	192	208
	124	140	156	175	195	216	262	312	348	431	64	80	90	101	112	124	134	156	171	192	215	234
	148	161	190	201	222	247	284	348	397	242	70	71	91	108	115	129	144	154	178	193	207	265
	165	187	207	230	257	285	345	416	54	66	81	98	110	120	134	147	162	185	198	223	257	280
	175	195	216	238	264	290	47	55	61	75	90	116	131	146	163	182	199	208	244	252	290	310
	198	201	228	252	280	41	50	60	65	80	96	123	143	159	176	196	211	232	262	281	321	350
	214	244	30	36	40	45	54	65	72	89	110	135	156	174	193	220	241	272	292	326	374	404
	231	31	35	39	43	48	57	69	77	94	116	144	161	178	197	225	245	275	300	330	380	420
	30	34	38	43	47	53	63	76	84	104	128	158	178	198	218	240	260	300	325	370	415	440
	34	39	43	48	53	59	71	85	94	115	142	175	197	220	243	270	285	320	340	390	450	485
	38	42	47	52	58	64	77	91	101	125	152	188	209	231	256	285	305	335	380	450	540	575
	40	45	51	57	62	69	84	101	112	137	169	206	232	261	289	290	315	355	480	485	565	590
	44	50	56	62	69	76	92	111	122	148	182	226	253	282	308	342	360	395	420	530	590	630
	49	55	61	68	76	84	102	121	133	161	197	246	277	308	342	360	410	430	455	580	605	660
	53	60	67	74	82	91	112	132	145	176	214	266	298	329	362	385	410	450	480	605	625	700
	58	66	74	82	91	102	121	142	158	192	235	289	324	358	396	410	440	490	515	630	700	745
	64	72	81	90	102	113	132	155	173	211	260	317	360	388	422	450	475	550	580	675	790	850
	69	78	87	97	107	118	141	165	183	225	279	338	374	414	458	495	540	620	670	750	865	935
	74	84	94	106	116	127	151	178	197	242	299	360	400	441	483	530	570	660	720	800	920	1005
	81	89	100	113	126	138	163	192	209	258	319	382	440	490	540	590	630	710	775	865	1000	1120
	86	96	108	121	133	146	172	202	221	272	337	404	461	508	562	610	665	740	810	905	1040	1120
	91	106	117	127	142	155	185	216	240	288	361	432	473	518	570	625	670	765	835	925	1065	1155
102	114	125	138	151	166	198	232	254	308	380	450	505	562	620	665	725	825	890	1000	1140	1230
104	116	126	141	155	170	200	237	258	313	380	465	520	578	632	700	740	850	925	1040	1200	1285
	117	130	143	157	172	190	224	263	289	352	417	500	575	608	680	765	810	940	1025	1130	1310	
	121	135	150	165	181	198	237	277	307	366	440	531	600	648	730	810	860	980	1165	1185		
	127	141	157	176	187	207	247	289	322	382	457	566	634	681	764	840	900	1040	1120			
	132	148	164	180	198	218	258	303	342	404	482	589	664	727	802	870	940					
	141	157	175	192	211	231	273	318	347	410	488	598	691	742	841	910						
	152	168	186	204	223	245	289	337	375	442	527	650	738	808	884							
	158	177	197	216	236	260	305	360	390	465	550	680	780	852								
	173	192	213	232	252	275	325	375	410	485	580	715	815	845								
	184	206	230	253	274	300	360	405	445	530	630	675	705	820	225	245	260	285	310	340	405	430
	196	218	238	258	280	310	375	430	475	560	635	670	210	220	235	245	270	285	315	340	375	420
	203	227	252	278	301	330	395	445	495	570	625	170	220	250	260	280	300	330	360	390	445	470
	211	236	261	287	315	340	410	470	510		650	640	235	265	270	295	310	345	375	410	485	515
	222	247	276	307	332	360	435	490	540	175	210	250	275	280	310	325	360	390	430	485	510	540
	230	256	288	318	352	380	470	520		185	220	260	290	295	320	340	380	410	450	510	540	
	248	274	307	340	376	415	500	500	160	195	230	275	305	310	335	355	395	425	470	530	560	
	254	284	317	351	390	430		510	165	200	245	285	315	320	350	370	430	470	490	560	585	
	261	293	325	362	400	445		140	170	210	250	280	330	335	340	365	390	430	470	510	580	610
	267	300	332	367	408		130	145	180	220	265	315	345	350	380	410	450	485	530	630	640	
	283	316	353	394			140	155	190	230	280	335	360	370	390	425	470	510	550	630	670	
	292	327	363				120	142	160	196	240	295	345	370	380	415	440	485	530	575	650	690
	310	346					126	151	168	209	255	300	365	385	400	435	460	510	550	580	680	720
	324					108	132	158	174	219	265	320	380	395	415	450	480	530	570	625	710	750
					104	114	138	166	184	230	280	335	390	415	435	470	500	560	605	660	740	795
			97	107	120	145	172	191	235	290	350	400	425	450	490	520	580	620	680	765	820	
			102	113	124	150	180	200	245	305	370	410	440	470	515	550	610	650	720	830	895	
		92	105	114	128	155	188	210	260	320	380	455	465	495	530	585	625	670	730	830	895	
		88	99	112	124	135	164	197	215	270	330	395	470	500	550	585	650	700	765	870	925	
	87	93	103	114	127	140	170	205	225	280	345	405	470	480	520	570	600	670	725	790	900	995
	91	97	108	120	133	147	179	213	235	290	360	420	480	500	545	590	630	695	745	815	950	1045
	100	112	123	135	147	161	190	220	250	300	375	435	490	520	555	610	655	745	805	895	1020	1080
	105	117	128	141	154	168	198	230	260	305	390	455	490	535	585	640	685	775	840	940	1120	1215
	110	122	134	147	161	175	210	240	270	315	405	470	510	565	615	665	720	820	880	990	1120	1215
	116	127	141	154	168	185	220	250	280	325	415	485	525	575	635	690	735	830	920	1000	1180	1275
	122	134	147	161	176	190	225	265	295	350	430	500	550	610	665	720	775	870	935	1045	1180	1275
	129	143	157	172	187	205	240	280	310	365	450	520	565	635	695	760	810	920				
	182	199	217	234	253	275	315	360	400	455	535	690	710	745	810							
	198	217	236	255	275	295	340	390		490	565	720	745	805								
10	1.15	1.20	1.25	1.30	1.35	1.40	1.50	1.60	1.66	1.79	1.93	2.10	2.20	2.28	2.37	2.46	2.54	2.68	2.76	2.89	3.05	3.15

3.45	3.60	3.74	3.95	4.15	4.40	4.59	4.73	5.18	5.40	5.77	6.07	6.21	6.45	6.86	7.08	8.34	9.89	11.9	Element
0.90	0.95	1.00	1.15	1.25	1.40	1.60	1.65	2.10	2.30	2.75	3.25	350	390	460	505	785	1250	22	1 H
2.25	2.55	2.85	3.30	3.85	4.60	5.15	5.60	7.35	8.20	10	11.50	12	13	1550	17	29	48	85	2 He
6.40	7.15	8.00	9.25	10.50	12	13	135	16	175	28	33	35.5	40	47	52	84	140	245	3 Li
13.5	15	16	17.5	25	31	35	38	50	56	68	79	85	96	114	127	200	330	565	4 Be
26	29	32	37	43	51	58	63	82	90	114	131	139	153	176	210	330	545	930	5 B
48	54	61	72	85	103	116	125	159	175	220	255	270	305	345	400	640	1050	1780	6 C
81	91	103	125	140	165	190	200	260	290	350	410	440	490	590	645	1040	1680	2800	7 N
123	142	162	180	210	245	275	300	385	435	530	615	650	735	875	965	1520	2440	4000	8 O
165	185	210	245	280	330	370	410	530	600	720	840	900	1015	1205	1315	1970	3110	5030	9 F
240	265	300	350	400	475	535	575	745	840	1015	1165	1230	1390	1605	1745	2700	4220	6640	10 Ne
294	330	366	430	490	580	650	715	915	1030	1230	1435	1525	1680	1940	2100	3395	4925	8160	11 Na
390	435	490	565	650	760	850	930	1120	1330	1580	1825	1962	2100	2450	2660	4050	350	590	12 Mg
480	535	595	690	790	915	1035	1130	1425	1610	1930	2165	2280	2500	2900	3170	330	500	850	13 Al
591	665	735	950	965	1125	1265	1375	1730	1960	2260	2565	2725	3000	290	315	480	740	1230	14 Si
685	770	850	970	1115	1290	1435	1570	1965	2180	2550	290	310	335	395	435	650	1015	1640	15 P
820	920	1030	1175	1375	1580	170	1920	220	250	300	345	365	410	480	525	795	1320	2100	16 S
923	1030	1140	1315	1480	170	190	210	275	300	365	420	450	490	585	635	960	1570	2500	17 Cl
1020	1144	1270	149	170	200	230	250	325	360	425	500	530	585	690	765	1160	1860	3000	18 Ar
132	147	162	186	215	250	280	305	380	425	500	575	610	670	780	855	1300	2120	3425	19 K
150	168	185	215	245	280	320	345	435	480	575	660	760	770	900	980	1500	2380	3850	20 Ca
174	195	215	250	285	330	370	400	505	560	665	765	810	890	1045	1135	1750	2680	4260	21 Sc
202	225	250	285	325	380	425	455	580	645	765	880	930	1020	1200	1300	2000	2975	4680	22 Ti
228	255	280	325	370	430	480	520	660	730	865	990	1050	1150	1340	1460	2200	3260	5050	23 V
263	295	320	375	425	490	550	590	755	835	985	1130	1200	1295	1540	1670	2470	3510	5480	24 Cr
297	330	360	420	475	550	675	665	840	935	1110	1270	1350	1460	1740	1920	2700	3790	5895	25 Mn
335	375	410	475	540	630	700	760	950	1070	1255	1450	1480	1620	1915	2040	2910	4100	6275	26 Fe
371	390	435	495	570	660	740	790	1040	1160	1350	1540	1590	1740	2030	2195	3070	4380	6640	27 Co
410	450	490	560	630	740	830	900	1150	1260	1470	1640	1710	1825	1970	2225	3140	4540	6900	28 Ni
450	495	530	610	690	795	885	960	1190	1350	1560	1760	1800	1950	2245	2415	3450	5035	7650	29 Cu
510	575	595	675	820	940	1050	1130	1310	1460	1670	1890	1940	2040	2325	2510	3645	5235		30 Zn
545	670	700	790	845	1010	1145	1225	1410	1575	1770	2000	2050	2185	2470	2645	3810			31 Ga
595	730	760	840	920	1095	1210	1320	1500	1670	1895	2110	2180	2300	2575	2750	3995			32 Ge
640	800	820	900	990	1165	1310	1420	1580	1795	2000	2225	2300	2400	2680	2880	1020	1580		33 As
700	868	925	1010	1100	1225	1380	1530	1710	1930	2135	2370	2440	2540	2800	3010	1110	1740		34 Se
760	950	1025	1100	1180	1305	1410	1615	1810	2060	2265	2495	2560	2680	2925	840	1190	1875		35 Br
820	1020	1100	1190	1280	1390	1505	1720	1940	2190	2410	2620	2700	2790	895	1300	2035			36 Kr
880	1090	1190	1290	1385	1495	1630	1810	2040	2330	2520			710	950	1380	2200			37 Rb
950	1150	1260	1400	1500	1615	1740	1910	2170	2465		655	720	760	1020	1500	2400			38 Sr
1025	1200	1310	1460	1620	1710	1875	2020	2290		635	660	695	770	810	1080	1620	2555		39 Y
1080	1260	1390	1565	1740	1880	1940	2125		615	675	720	755	800	875	930	1230	1870	2755	40 Zr
1180	1310	1450	1660	1870	2140			590	655	720	755	805	850	935	990	1315	2005	3180	41 Nb
1260	1400	1540	1775	2000			575	630	700	770	805	850	935	990	1315	2005	3180		42 Mo
1370	1510	1670	1880			485	610	670	745	825	860	905	1005	1060	1410	2160	3420		43 Tc
1425	1570	1720		460	490	525	560	705	785	870	900	950	1055	1120	1490	2280	3620		44 Ru
1440	1590		410	490	520	560	590	690	750	835	920	958	1015	1120	1180	1580	2435	3840	45 Rh
		385	435	520	550	590	590	800	860	945	980	1020	1075	1190	1260	1675	2590	4100	46 Pd
1450	355	408	460	550	585	625	790	860	945	1040	1075	1160	1280	1350	1800	2700	4230		47 Ag
	362	385	435	490	585	620	665	890	890	995	1100	1140	1210	1340	1415	1880	2930	4560	48 Cd
340	380	400	455	515	610	650	700	860	930	1040	1150	1200	1260	1340	1480	1975	3090	4800	49 In
365	410	435	490	550	660	720	760	915	990	1100	1210	1260	1340	1475	1575	2280	3360	5300	50 Sn
305	425	455	510	585	680	730	780	960	1050	1160	1280	1340	1420	1560	1660	2355	3500	5510	51 Sb
405	445	480	535	605	720	770	825	1005	1100	1225	1350	1400	1480	1640	1740	2500	3650	5625	52 Te
425	470	505	565	640	760	805	870	1060	1160	1290	1425	1480	1570	1740	1840	2645	3870		53 I
450	495	530	595	670	800	850	910	1120	1220	1330	1485	1560	1660	1830	1940	2790			54 Xe
475	525	560	625	700	845	895	960	1180	1280	1420	1575	1640	1730	1915	2010				55 Cs
495	545	585	660	745	880	940	1010	1230	1345	1490	1640	1720	1825	2005	2130				56 Ba
520	570	610	690	780	925	980	1055	1290	1400	1560	1720	1800	1900	2090	2210				57 La
540	600	640	720	815	965	1025	1110	1355	1475	1630	1800	1885	2000	2195					58 Ce
570	630	675	760	850	1015	1070	1160	1420	1550	1710	1880	1975	2095						59 Pr
595	660	710	795	890	1060	1125	1210	1475	1610	1780	1970	2070	2180						60 Nd
625	695	745	835	935	1110	1180	1275	1550	1675	1860	2060	2155							61 Pm
655	720	780	865	995	1160	1230	1320	1620	1760	1950	2155								62 Sm
680	755	810	905	1015	1205	1280	1380	1680	1840	2010									63 Eu
715	790	845	945	1060	1260	1345	1445	1760	1910	2110									64 Gd
740	825	885	985	1110	1320	1400	1510	1840	2000										65 Tb
780	860	920	1035	1160	1380	1460	1560	1920	2090										66 Dy
815	895	950	1070	1220	1440	1525	1640	2000											67 Ho
840	930	1000	1120	1260	1480	1580	1690	2175											68 Er
880	970	1040	1170	1310	1550	1650	1760												69 Tm
920	1020	1080	1210	1370	1610	1720	1840												70 Yb
960	1050	1140	1260	1420	1680	1750													71 Lu
990	1070	1170	1320	1480	1730	1815													72 Hf
1035	1130	1230	1370	1530	1810														73 Ta
1080	1180	1270	1420	1595															74 W
1125	1230	1325	1475																75 Re
1160	1275	1375	1630																76 Os
1210	1330	1425																	77 Ir
1370	1500																		78 Pt
1420	1575																		79 Au
1505																			80 Hg
1510																			81 Tl
1570																			82 Pb
																			83 Bi
																			90 Th
																			92 U

| 3.45 | 3.60 | 3.74 | 3.95 | 4.15 | 4.40 | 4.59 | 4.73 | 5.18 | 5.40 | 5.77 | 6.07 | 6.21 | 6.45 | 6.86 | 7.08 | 8.34 | 9.89 | 11.9 |

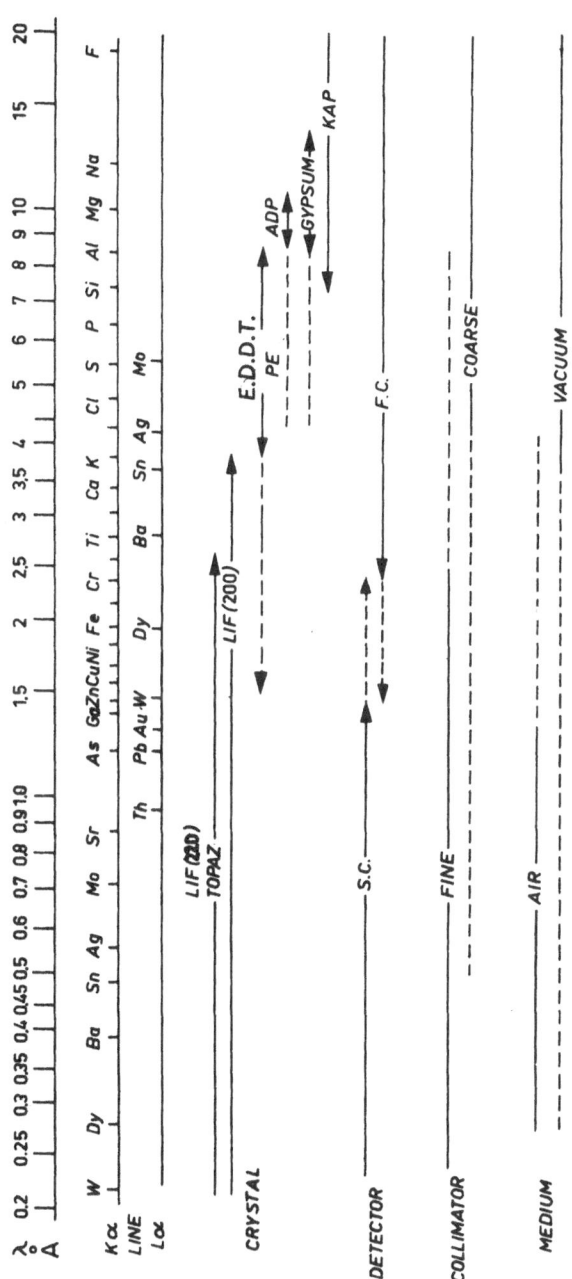

APPENDIX 2(a) Choice of conditions for X-ray fluorescence analysis.

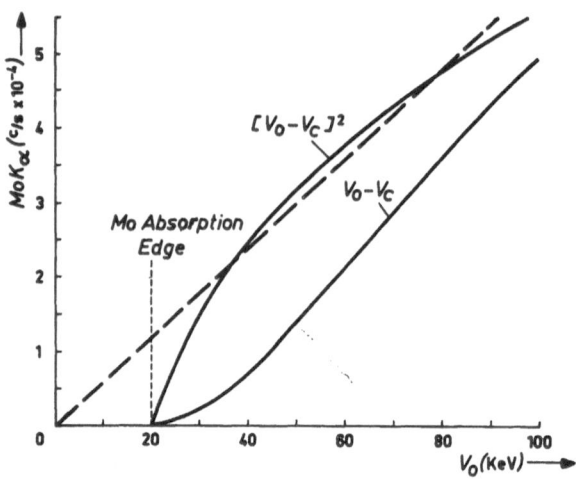

APPENDIX 2(b) Variation of characteristic line intensity with X-ray tube voltage
(a) $I \propto V_O$ ... where V_O is 2 to 5 times V_c
(b) $I \propto V_O\text{-}V_c$ ... where V_O is $> 2\ V_c$
(c) $I \propto (V_O\text{-}V_c)^2$ ... where V_O is up to $2\ V_c$

Note: $I \propto$ tube current.
Appendix 2(b)

TABLE I. PRINCIPAL EMISSION LINES OF X-RAY SPECTRA (EMISSION WAVELENGTHS IN Å UNITS)

K Series (Li–Ru)

Line	Transition K, Approx. intensity (Rel. Kα₁)	α° (150)	α₁ (100)	α₂ (50)	β₁ Mₙₙ (15)	β₃ Mₘ	β₂ NₙNₙ (5)	β₄ NᵢᵥNᵥ (<1)	β₅ MᵢᵥMᵥ (<1)	— Oₙₙₙ (<1)	K Absorption Edge
Li	3	240.									226.953
Be	4	113.									
B	5	67.									
C	6	44.									43.767
N	7	31.603									31.052
O	8	23.707									23.367
F	9	18.307									
Na	11	11.909			11.617						
Mg	12	9.889			9.558						9.512
Al	13	8.339	8.338	8.341	7.981						7.951
Si	14	7.126	7.125	7.127	6.769						6.744
P	15	6.155			5.804						5.787
S	16	5.373	5.372	5.375	5.032						5.018
Cl	17	4.729	4.728	4.731	4.403						4.397
A	18	4.192	4.191	4.194						3.442	3.871
K	19	3.744	3.742	3.745	3.454	3.442					3.437
Ca	20	3.360	3.359	3.362	3.089	3.074					3.070
Sc	21	3.034	3.031	3.034	2.786	2.764					2.758
Ti	22	2.750	2.749	2.753	2.514	2.498					2.497
V	23	2.505	2.503	2.507	2.285	2.270					2.269
Cr	24	2.291	2.290	2.294	2.085	2.071					2.070
Mn	25	2.103	2.102	2.105	1.910	1.897					1.897
Fe	26	1.937	1.936	1.940	1.757	1.745					1.744
Co	27	1.791	1.789	1.793	1.621	1.609					1.608
Ni	28	1.659	1.658	1.661	1.500	1.489					1.488
Cu	29	1.542	1.540	1.544	1.392	1.382	1.381				1.381
Zn	30	1.437	1.435	1.439	1.296	1.284	1.284				1.281
Ga	31	1.341	1.340	1.344	1.208	1.207	1.196				1.195
Ge	32	1.256	1.255	1.258	1.129	1.129	1.117				1.116
As	33	1.177	1.175	1.179	1.058	1.057	1.045				1.045
Se	34	1.106	1.105	1.109	0.993	0.992	0.980				0.980
Br	35	1.041	1.040	1.044	0.933	0.933	0.921	0.866			0.920
Kr	36	0.981	0.980	0.984	0.879	0.879	0.866	0.866			0.866
Rb	37	0.927	0.926	0.930	0.830	0.830	0.817				0.816
Sr	38	0.877	0.875	0.880	0.784	0.783	0.771				0.770
Y	39	0.831	0.829	0.833	0.741	0.740	0.728	0.727			0.727
Zr	40	0.788	0.786	0.791	0.702	0.701	0.690				0.688
Cb	41	0.748	0.747	0.751	0.666	0.665	0.654				0.653
Mo	42	0.710	0.709	0.713	0.632	0.632	0.631	0.620			0.620
Tc	43	0.674	0.673	0.676	0.602	0.602					
Ru	44	0.644	0.643	0.647	0.573	0.573	0.562				0.560

K Series (Rh–U)

Line	Transition K, Approx. intensity (Rel. Kα₁)	α° (150)	α₁ (100)	α₂ (50)	β₁ Mₙₙ (15)	β₃ Mₘ (15)	β₂ NₙVₙₙ (s)	β₄ NₙNₙ (<1)	β₅ MᵢᵥMᵥ	— Oₙₙₙ	K Absorption Edge
Rh	45	0.614	0.613	0.617	0.546	0.546	0.535				0.534
Pd	46	0.587	0.585	0.590	0.521	0.521	0.510				0.509
Ag	47	0.561	0.559	0.564	0.497	0.498	0.487				0.486
Cd	48	0.536	0.535	0.539	0.475	0.476	0.465				0.464
In	49	0.514	0.512	0.517	0.455	0.455	0.445				0.444
Sn	50	0.492	0.491	0.495	0.435	0.436	0.426				0.425
Sb	51	0.472	0.470	0.475	0.417	0.418	0.408				0.407
Te	52	0.453	0.451	0.456	0.401	0.401	0.391				0.390
I	53	0.435	0.433	0.438	0.384	0.385	0.376				0.374
X	54	0.418	0.416	0.421	0.369	0.355	0.360				0.359
Cs	55	0.402	0.401	0.405	0.355	0.355	0.346				0.345
Ba	56	0.387	0.385	0.390	0.341	0.342	0.333				0.332
La	57	0.373	0.371	0.376	0.328	0.329	0.320				0.319
Ce	58	0.359	0.357	0.362	0.316	0.317	0.309				0.307
Pr	59	0.346	0.344	0.349	0.305	0.305	0.297				0.296
Nd	60	0.334	0.332	0.337	0.294	0.294	0.287				0.285
Il	61	0.322	0.321	0.325	0.283						
Sm	62	0.311	0.309	0.314	0.274	0.274	0.267				0.265
Eu	63	0.301	0.299	0.304	0.264	0.265	0.258				0.256
Gd	64	0.291	0.289	0.294	0.255	0.256	0.249				0.246
Tb	65	0.281	0.280	0.284	0.246	0.238	0.239				0.238
Dy	66	0.272	0.270	0.275	0.237	0.238	0.231				0.230
Ho	67	0.263	0.261	0.266	0.223	0.223		0.217			0.223
Er	68	0.255	0.253	0.258	0.215	0.216	0.217				0.215
Tu	69	0.246	0.244	0.250	0.208	0.209	0.203				0.209
Yb	70	0.238	0.236	0.241	0.202	0.203	0.197	0.203			0.202
Lu	71	0.231	0.229	0.234	0.190	0.196	0.185	0.197			0.195
Hf	72	0.224	0.222	0.227	0.195	0.196	0.190	0.190			0.189
Ta	73	0.217	0.215	0.220	0.190	0.191	0.185	0.185			0.184
W	74	0.211	0.209	0.213	0.184	0.185	0.179	0.179			0.178
Re	75	0.204	0.202	0.207	0.173	0.179	0.174	0.174			0.173
Os	76	0.198	0.196	0.201	0.169	0.173	0.169	0.169			0.168
Ir	77	0.193	0.191	0.196	0.167	0.168	0.164	0.164	0.163	0.162	0.163
Pt	78	0.185	0.185	0.190	0.163	0.164	0.160	0.159	0.158	0.162	0.158
Au	79	0.182	0.180	0.185	0.159	0.160	0.155	0.154	0.158	0.158	0.153
Hg	80										0.149
Tl	81	0.172	0.170	0.175	0.150	0.151	0.147		0.147	0.145	0.144
Pb	82	0.167	0.165	0.170	0.146	0.143	0.138		0.141		0.141
Bi	83	0.162	0.161	0.165	0.142	0.118	0.114			0.116	0.137
Th	90	0.135	0.133	0.138	0.117	0.118	0.114			0.113	0.115
U	92	0.128	0.126	0.131	0.111	0.112	0.108				0.107

*Kα° ... Unresolved Kα₁ Kα₂

Table II. Principal Emission Lines of X-ray Spectra (Emission Wavelengths in Å Units)

L Series

Element	Z (Int. Lα)	Lα₁	Lα₂	Lβ₃	Lβ₁	Lβ₄	Lγ₁	Lγ₂	Lγ₃	Lℓ	Lη	L	L_II	L_III
Cl	17									67.64	67.25			
A	18									56.312	56.813			
K	19									47.885	47.325			
Ca	20	36.393	36.022							41.042	40.542		42.164	
Sc	21	31.993	31.072							35.671	35.200		35.500	35.561
Ti	22	27.445	27.074							31.423	30.943		27.39	
V	23	24.309	23.696							27.686	27.375			
Cr	24	21.713	21.323							24.940	24.339	16.7	17.9	20.7
Mn	25	19.489	19.158							22.315	21.864			
Fe	26	17.602	17.290							20.201	19.73			
Co	27	16.000	15.698							18.358	17.86			
Ni	28	14.595	14.308							16.693	16.304			
Cu	29	13.357	13.079							15.297	14.940			
Zn	30	12.282	12.059							14.081	13.719	13.010	13.299	
Ga	31	11.313	11.065							12.976	12.620	11.861	12.130	
Ge	32	10.456	10.194							11.944	11.606			
As	33	9.671	9.414							11.069	10.732	8.108	9.124	9.367
Se	34	8.990	8.735							10.293	9.959	7.506	8.417	8.645
Br	35	8.375	8.126							9.583	9.253			
Kr	36													
Rb	37	7.318	7.075	6.788	6.821	6.994	6.754	6.045		8.363	8.042	5.997	6.643	6.864
Sr	38	6.863	6.623	6.367	6.403	6.519	6.297	5.644		7.836	7.517	5.582	6.172	6.387
Y	39	6.449	6.211	5.983	6.018	6.094	5.875	5.283	5.384	7.356	7.040	5.233	5.756	5.962
Zr	40	6.070	5.836	5.632	5.666	5.710	5.497	4.953		6.918	6.606	4.867	5.378	5.583
Cb	41	5.725	5.492	5.310	5.346	5.361	5.151	4.654	5.036	6.517	6.210	4.581		5.225
Mo	42	5.406	5.176	5.013	5.048	5.048	4.837	4.390	4.726	6.150	5.847	4.299	4.719	4.913
Ma	43													
Ru	44	4.846	4.620	4.487	4.523	4.487	4.288	3.897	4.182	5.503	5.204	3.636	4.179	4.369
Rh	45	4.597	4.374	4.253	4.289	4.242	4.045	3.685	3.944	5.217	4.922		3.942	4.139
Pd	46	4.368	4.146	4.034	4.071	4.016	3.822	3.489	3.725	4.953	4.660	3.428	3.724	3.908
Ag	47	4.154	3.935	3.834	3.870	3.808	3.616	3.307	3.523	4.707	4.418	3.254	3.514	3.698
Cd	48	3.956	3.739	3.644	3.661	3.614	3.426	3.137	3.336	4.480	4.193	3.084	3.326	3.504
In	49	3.753	3.555	3.470	3.507	3.436	3.249	2.980	3.163	4.269	3.983	2.926	3.147	3.355

Sn 50	3.156	2.982	2.778		3.789	4.071		3.085	2.778	2.835	3.001		3.121	3.115	3.155	3.170	3.344	3.306	3.175	3.385	3.609	3.600
Sb 51	3.000	2.830	2.639		3.607	3.888		2.932	2.639	2.695	2.852		2.979	2.973	3.005	3.115	3.190	3.152	3.023	3.226	3.448	3.439
Te 52	2.856	2.687	2.510		3.438	3.716		2.790	2.511	2.567	2.712		2.847	2.839	2.863	2.971	3.046	3.009	2.882	3.077	3.299	3.290
I 53	2.719	2.553	2.389		3.280	3.557		2.657	2.391	2.447	2.582		2.720	2.713	2.730	2.837	2.912	2.874	2.751	2.937	3.157	3.148
X 54	2.592	2.429	2.274																			
Cs 55	2.474	2.314	2.167		2.994	3.267	2.222	2.417	2.174	2.237	2.348		2.492	2.485	2.478	2.593	2.666	2.628	2.511	2.663	2.902	2.892
Ba 56	2.363	2.204	2.068		2.862	3.135		2.309	2.075	2.134	2.243		2.387	2.376	2.382	2.482	2.555	2.516	2.404	2.567	2.785	2.776
La 57	2.259	2.103	1.973		2.740	3.006		2.205	1.983	2.041	2.141		2.290	2.282	2.275	2.379	2.449	2.410	2.303	2.458	2.674	2.665
Ce 58	2.164		1.890		2.620	2.892	2.023	2.110	1.899	1.955	2.048		2.195	2.188	2.180	2.282	2.349	2.311	2.208	2.356	2.570	2.561
Pr 59	2.077	1.924	1.811		2.512	2.784	1.956	2.020	1.819	1.874	1.961		2.107	2.100	2.091	2.190	2.255	2.216	2.119	2.259	2.473	2.463
Nd 60	1.995	1.843	1.735		2.409	2.675		1.935	1.745	1.797	1.878		2.023	2.016	2.009	2.103	2.166	2.126	2.035	2.166	2.382	2.370
Il 61							1.855												2.081			2.283
Sm 62	1.845	1.702	1.598	1.831	2.218	2.482	1.632	1.708	1.606	1.655	1.726	1.779	1.870	1.862	1.856	1.946	2.000	1.962	1.882	1.998	2.210	2.199
Eu 63	1.776	1.626	1.536	1.776		2.395			1.597	1.591	1.657		1.800	1.792	1.788	1.875	1.926	1.887	1.812	1.920	2.131	2.120
Gd 64	1.709	1.561	1.477		2.049	2.312			1.485	1.534	1.592		1.731		1.723	1.807	1.853	1.815	1.746	1.847	2.057	2.046
Tb 65	1.648	1.501	1.421	1.663		2.234		1.544	1.427	1.471	1.530	1.577	1.667		1.659	1.742	1.785	1.747	1.682	1.777	1.986	1.976
Dy 66	1.579	1.438	1.365	1.612	1.898	2.158		1.374	1.417	1.471	1.473			1.599	1.599	1.681	1.720	1.681	1.623	1.710	1.920	1.909
Ho 67	1.535	1.390	1.318		1.826	2.086		1.462	1.323	1.364	1.417		1.494			1.622	1.658	1.619	1.567	1.647	1.856	1.845
Er 68	1.482	1.339	1.269		1.757	2.019		1.406	1.276	1.315	1.364	1.494		1.494		1.567	1.601	1.561	1.514	1.587	1.796	1.785
Tu 69	1.433	1.288	1.222		1.695	1.955		1.355		1.268	1.316			1.515	1.515	1.544	1.505	1.463	1.530	1.738	1.726
Yb 70	1.386	1.243	1.181		1.635	1.894		1.307	1.185	1.222	1.268	1.392	1.387	1.395	1.466	1.491	1.452	1.416	1.476	1.682	1.672
Lu 71	1.342	1.198	1.140		1.478	1.836		1.260	1.143	1.179	1.222	1.372	1.343	1.384	1.350	1.419	1.441	1.402	1.370	1.424	1.630	1.619
Hf 72	1.298	1.154	1.099		1.523	1.782		1.215	1.103	1.138	1.179	1.328	1.299	1.306	1.374	1.392	1.353	1.327	1.374	1.580	1.569
Ta 73	1.256	1.113	1.061		1.471	1.728		1.173	1.065	1.105	1.138	1.287	1.254	1.264	1.247	1.290	1.346	1.307	1.285	1.327	1.533	1.522
W 74	1.215	1.074	1.024		1.421	1.678		1.132	1.028	1.062	1.098	1.247	1.212	1.204	1.224	1.290	1.302	1.263	1.245	1.282	1.487	1.476
Re 75	1.177	1.037	.990		1.374	1.630	1.044	1.094	.993	1.026	1.061	1.206	1.172	1.165	1.186	1.252	1.260	1.220	1.206	1.238	1.444	1.433
Os 76	1.140	1.001	.956		1.328	1.585	1.008	1.057	.959	.992	1.025	1.171	1.133	1.126	1.149	1.213	1.218	1.179	1.169	1.197	1.402	1.391
Ir 77	1.105	.967	.923		1.285	1.541	.974	1.022	.928	.959	.991	1.137	1.097	1.090	1.115	1.179	1.179	1.141	1.135	1.158	1.363	1.352
Pt 78	1.072	.934	.893		1.243	1.499	.941	.988	.897	.928	.958		1.062	1.054	1.082	1.143	1.142	1.104	1.102	1.120	1.325	1.313
Au 79	1.040	.903	.864	1.414	1.202	1.460	.910	.956	.867	.898	.927	1.128	1.072	1.021	1.050	1.111	1.106	1.070	1.083	1.083	1.288	1.277
Hg 80	1.009	.872	.836		1.164	1.422	.880	.925	.839	.869	.897	1.090	1.019	1.028	1.060	1.072	1.034	1.040	1.049	1.253	1.242
Tl 81	.979	.844	.808	1.342	1.127	1.385	.852	.895	.812	.842	.868	1.012	.996	.957	.990	1.039	1.001	1.001	1.015	1.218	1.207
Pb 82	.950	.815	.782	1.308	1.092	1.350	.824	.867	.761	.822	.840	1.022	.964	.934	.962	1.021	1.007	.969	.983	1.186	1.175
Bi 83	.924	.789	.757	1.210	1.058	1.317	.799	.840	.761	.790	.814	.989	.905	.898	.935	.993	.977	.939	.955	1.155	1.144
Po 84										.788		.931	.900	.967		.948	.909	.929	1.125	1.114
Fr 87	.803	.670	.644		.908	1.167	.680	.717	.649	.675	.694	.844	.838	.776	.817	.871	.841	.803	.836	.814	1.017	1.005
Ra 88	.761	.630	.606	1.080	.855	1.115	.640	.675	.611	.635	.653		.730	.775	.828	.793	.755	.766	.958	.956
Th 90					.830	1.091	.601	.655	.594	.617	.634		.708	.701	.755	.770	.732	.794	.774	.945	.933
Pa 91						1.067		.635	.613		.615	.736	.681	.806	.755	.710	.720	.923	.911
U 92	.722	.592	.569	1.035	.806		.601	.635	.577	.598	.605		.687	.698	.735	.748	.726	.698	.889	
Np 93											.597								

APPENDIX 4

DERIVATION OF EQUATIONS FOR FIXED TIME OPTIMAL AND FIXED COUNT

Fixed time optimal

The ratio $\dfrac{T_p}{T_b} = a$ is calculated to give a minimum error in the measurement $R_p - R_b$ in a total time T. This is the value of a for which $\dfrac{d\sigma}{da} = 0$. Substitution of $T_p = \dfrac{aT}{1+a}$ and $T_b = \dfrac{T}{1+a}$ in equation (5.10)

$$\sigma = \sqrt{\frac{(1+a)R_p}{aT} + \frac{(1+a)R_b}{T}}$$

Calculation of $\dfrac{d\sigma}{da} = 0$ will reduce itself to solving the equation.

$$\frac{d\sigma}{da}\left[\left(\frac{1+a}{a}\right)R_p + (1+a)R_b\right] = 0$$

or $\dfrac{a-(1+a)}{a^2}R_p + R_b = 0$

which gives $a^2 = \dfrac{R_p}{R_b}$ and $a = \sqrt{\dfrac{R_p}{R_b}}$ \hfill (5.14)

Substitution of (5.14) in (5.10) gives the minimum obtainable σ in a time T for given values of R_p and R_b.

$$\sigma_{F.T.O.} = \frac{1}{\sqrt{T}} \cdot \sqrt{\left(1 + \sqrt{\frac{R_p}{R_b}}\right)\left(\frac{R_p}{R_b}\right)^{-\frac{1}{2}} \cdot R_p\left(1 + \sqrt{\frac{R_p}{R_b}}\right)R_b}$$

$$\therefore \sigma_{F.T.O.} = \frac{1}{\sqrt{T}}\sqrt{R_p + \sqrt{2R_pR_b} + R_b}$$

$$\therefore \sigma_{F.T.O.} = \frac{1}{\sqrt{T}}\left(\sqrt{R_p} + \sqrt{R_b}\right) \hfill (5.15)$$

$$\varepsilon\%_{F.T.O.} = \frac{100}{\sqrt{T}}\frac{\sqrt{R_p} + \sqrt{R_b}}{R_p - R_b}$$

$$\therefore \varepsilon\%_{F\ T.O.} = \frac{100}{\sqrt{T}} \cdot \frac{1}{\sqrt{R_p} - \sqrt{R_b}} \hfill (5.16)$$

Fixed count

The formula given for the fixed count method is obtained by expression R_p and T_b in T, R_p and R_s.

From $R_p T_p = R_b T_b = N$

follows $\dfrac{T_p}{T_b} = \dfrac{R_b}{R_p}$

substitution in $T_p + T_b = T$ gives

$$T_b = \frac{R_p T}{R_p + R_b}$$

$$T_p = \frac{R_b T}{R_p + R_b}$$

substitution in (5.10) gives

$$\sigma_{F.C.} = \sqrt{\frac{R_p(R_p + R_b)}{R_b T} + \frac{R_b(R_p + R_b)}{R_p T}}$$

$$\sigma_{F.C.} = \frac{1}{\sqrt{T}} \sqrt{R_p + R_b} \sqrt{\frac{R_p}{R_b} + \frac{R_b}{R_p}} \tag{5.17}$$

INDEX